Artificial Intelligence Applications in Human Pathology

Artificial Intelligence Applications in Human Pathology

editors

Ralf Huss
University Hospital Augsburg, Germany

Michael Grunkin
Visiopharm, Denmark

World Scientific

NEW JERSEY · LONDON · SINGAPORE · BEIJING · SHANGHAI · HONG KONG · TAIPEI · CHENNAI · TOKYO

Published by

World Scientific Publishing Europe Ltd.

57 Shelton Street, Covent Garden, London WC2H 9HE

Head office: 5 Toh Tuck Link, Singapore 596224

USA office: 27 Warren Street, Suite 401-402, Hackensack, NJ 07601

Library of Congress Cataloging-in-Publication Data

Names: Huss, Ralf, editor. | Grunkin, Michael, editor.

Title: Artificial intelligence applications in human pathology / editors,
     Ralf Huss (University Hospital Augsburg, Germany), Michael Grunkin (Visiopharm, Denmark).

Description: New Jersey : World Scientific, [2022] | Includes bibliographical references and index.

Identifiers: LCCN 2021038679 | ISBN 9781800611382 (hardcover) |
     ISBN 9781800611399 (ebook for institutions) | ISBN 9781800611405 (ebook for individuals)

Subjects: LCSH: Diagnosis, Laboratory--Data processing. | Artificial intelligence--
     Medical applications. | Pathology--Technological innovations.

Classification: LCC RB38 .A78 2022 | DDC 616.07/5--dc23

LC record available at https://lccn.loc.gov/2021038679

British Library Cataloguing-in-Publication Data

A catalogue record for this book is available from the British Library.

For any available supplementary material, please visit
https://www.worldscientific.com/worldscibooks/10.1142/Q0336#t=suppl

Desk Editors: Jayanthi Muthuswamy/Michael Beale/Shi Ying Koe

Typeset by Stallion Press
Email: enquiries@stallionpress.com

Foreword

With this book, the *Artificial Intelligence (AI) Applied in Human Pathology* group illustrates and explores the expanding dimensions of a great new era in the application of digital pathology and AI. Although fresh, this book is long overdue, because these are "hot topics" on many pathology meetings around the world.

In the preface of his famous textbook, Stanley Robbins wrote: *The pathologist is interested not only in the recognition of structural alterations, but also in their significance, i.e., the effects of these changes on cellular and tissue function and ultimately the effect of these changes on the patient. It is not a discipline isolated from the living patient, but rather a basic approach to a better understanding of disease and therefore a foundation of sound clinical medicine.*

The public would be surprised to learn that pathology has not been a "high tech" medical imaging specialty like radiology. Observation and dissection have been its mainstays for decades, with photography, histology, and immunohistochemistry as the main tools, and molecular pathology with next generation sequencing being increasingly important in the last 10 years. Only in the past few years have several institutions around the world decided to digitize their entire pathology workflows. The possibility of digitizing whole-slide images of tissue has allowed to use AI in digital pathology.

The incorporation of digital pathology into daily operational pathology practice, however, promises to strengthen our discipline enormously. In combination with AI, perhaps, even to redefine it. It will not replace pathologists, but it will do two things: (1) it will help to share images and data, increase efficiency, integrate diagnostics, and lead to modernization of pathology workflows, and (2) by using AI algorithms many of the tasks that are manual and subjective can become more automated and standardized. "Computational Pathology" then will guide pathologists through new terrain which is familiar in one sense but quite foreign in another. I am convinced that computational pathology and AI can have immense potential for oncology and precision medicine. Of course, there are many challenges that need to be overcome in computational pathology, but this digital innovation has potential to change the way we diagnose cancer. Bringing the disciplines of pathology and AI together will help advance the future of precision oncology.

I am honored to have been asked by Ralf Huss and Michael Grunkin to write this foreword. Over many years, they have had the vision, initiative, and energy to build the knowledge base for the contribution of digital pathology, including AI, to general pathology. This book is timely and significant. The convergence of digital pathology with AI and personalized medicine driven by improved pathology diagnostics is revolutionizing the whole field.

Holger Moch, M.D.
Professor & Chairman
Department of Pathology and Molecular Pathology
University Hospital Zurich
President of the European Society of Pathology

About the Editors

Ralf Huss, M.D., Ph.D., is professor of pathology and managing director of the Institute of General Pathology and Molecular Diagnostics, University Hospital in Augsburg (Germany), where he also chairs the Institute for Digital Medicine. Dr. Huss is board certified in anatomical, experimental, and molecular pathology with more than 30 years of experience in histopathology, immunology, cancer research, and oncology. Previously, Dr. Huss held different senior positions in the pharmaceutical industry (Roche, AstraZeneca). He also serves as honorary professor at the University College Dublin (Ireland) and as lecturer of nanotechnology and biomaterials at the Faculty of Chemistry, Technical University of Munich (Germany). His research interest is mainly on the use of machine learning and artificial intelligence in routine tissue diagnostics and biomarker discovery.

Michael Grunkin, Ph.D., is co-founder and CEO of Visiopharm A/S, which is a Denmark-based world leader in AI-driven precision pathology and digital pathology for research and diagnostics. Before establishing Visiopharm in 2001, Dr. Grunkin was technical-founder and director of R&D in two other AI-driven medical device companies, specializing in diagnosis of osteoporosis and charactering progression of diabetic retinopathy, respectively. In 1993, Dr. Grunkin received his Ph.D. from the Technical University of Denmark (TUD), in the fields of applied mathematics, spatial statistics, and computer science applied in the life sciences. Over a period of two years (1993–1995), he was a visiting scientist and post-doc at Massachusetts Institute of Technology, where his research focused on applying AI/ML in the context of life science research and diagnostics. After that, he became an assistant professor at TUD where he founded and taught a Medical Image Analysis course, published several research papers and patents, and was advisor for numerous Master's Thesis projects. Dr. Grunkin continues to co-supervise Master's and Ph.D. students within this technical discipline.

Contents

https://doi.org/10.1142/9781800611399_0001

Chapter 1

Introduction: Integration of Computational Pathology and AI Application into Histopathology Workflow

Ralf Huss

*Institute of Pathology and Molecular Diagnostics,
Medical Faculty, University Augsburg, Augsburg, Germany*

*Institute for Digital Medicine,
University Hospital Augsburg, Augsburg, Germany*

ralf.huss@uk-augsburg.de

Until recently, the conventional role of the pathologist has been to present anatomical pathology findings and associated diagnosis, particularly as a written report or at interdisciplinary and molecular tumor boards. Recently, the responsibilities of pathologists have greatly expanded dealing with an increasing level of complexity and in-depth analytical capabilities. The sole knowledge base of histopathologic diagnoses that usually lead to therapeutic decisions is no longer sufficient in the emerging world of molecular and precision oncology, which holds too many different and complex inconclusive implications for distinct types and subtypes related to their diagnosis and treatment options on the individual patient level.

Accurate and timely diagnosis is critical for deciding on the best treatment for the individual patient at the right time. Pathologists have assumed a more central role and are again essential physicians, especially on the oncology care team, because they have specialized knowledge to inform diagnostic test selection, performance, and interpretation at the highest quality level, as well as the communication of results and the implications for subsequent care decisions — a role that pathologists self-confidently owned after the successful implementation of light microscopy about 200 years ago.

Artificial intelligence (AI) is another powerful technology like the light microscope, but many pathologists do not yet know how AI could be applied in their daily clinical work. Yet, pathologists must participate in the development and clinical use of AI-based solutions to deliver the digital future and guide the generation of diagnostic and predictive algorithms through their existing expert knowledge and visual experience. Without the profound knowledge of pathologists and their professional awareness of necessary high-quality standards in the processing of tissue and cytology specimens, a successful integration of AI-based solutions into a clinical pathology workflow is an endeavor doomed to failure.

Machine learning (ML) is a collection of mathematical and computer science techniques for the extraction of relevant data from large datasets trained and explained by human experts and in this case by expert pathologists. Therefore, ML can support the human factor, mostly in performing difficult, tedious, and repetitive tasks. An automated ML method will read multiplexed immunohistochemistry images always in the same reliable and robust manner, yielding the identical result over and over again. This allows a global comparability of complex and larger datasets without significant inter- and intra-observer variability. ML algorithms may include decision trees, Bayes's networks, clustering solutions, and regression solutions as well, depending on whether the solution is rather supervised or unsupervised.

With the advent of convolutional neural networks (CNNs) in recent years, ML is becoming increasingly accessible to researchers from non-computer science backgrounds. However, preparing data and images for ML remains a crucial step which requires hands-on expertise and expert knowledge.

It is "remarkable" how a computer can already "read" and "interpret" a digitized slide and usually "deliver" a more accurate and quantitative interpretation than just through eyeballed microscopic viewing.

Yet, machine-based decision tools either require a comprehensive training of the algorithms and many possible iterations using big sets of data or human experts (pathologists) to create heuristic rule sets. However, for the time being, and until a broader acceptance of computational pathology (CP) in the medical community, it remains also legally necessary to confirm any machine-supported diagnosis through a highly skilled and well-trained pathologist and to provide a plausibility check on the accuracy of the computer-assisted decision. This applies even more to any prediction and treatment recommendation, which also relies on the appropriate pre- and post-analytical quality of the tissue sample and the analytic process.

Advances in ML and AI in general have propelled CP research. Today, computer systems approach the diagnostic levels achieved by humans for certain well-defined tasks in pathology [1]. At the same time, pathologists are faced with an increased workload both quantitatively (numbers of cases) and qualitatively (the amount of work per case with increasing treatment options and the type of data delivered by pathologists is also expected to become more fine-grained).

CP will leverage mathematical tools and implement data-driven methods as a center for data interpretation in modern tissue diagnosis. The value proposition of CP, as well as digital pathology to some extent, will require working with institutional administrations, regulatory offices, notified bodies, other departments including informatics and mathematics, and of course colleagues in pathology. CP will also foster the training of future computational pathologists, those with both pathology and non-pathology backgrounds, who will eventually decide that CP will serve as an indispensable hub for data-related research in a global healthcare system.

Mathematical and theoretical (algebraic) pathology employs theoretical analysis, appropriate mathematical models, and abstractions to investigate the principles that govern the structure, development, and hidden information of tissue specimens. The field is close to the discipline of bioinformatics and aims at the mathematical representation, visualization, and modeling of pathological conditions, using techniques and tools of applied mathematics, informatics, and statistics. Discrete components of biological systems arise from state transitions (e.g., from healthy to abnormal/malignant tissues), abstractions and approximations, nonlinear effects, and the presence of inherently discrete processes, often observed in systems governed by one, few, and many different events in a single cell or a complex tissue. Here, applied mathematics can be useful in both theoretical and practical applications also in anatomical pathology. Algebraic topology including spatial modeling, swam behavior, pattern formation, etc., uses tools from abstract algebra to study topological spaces. The goal is to find algebraic invariants from all complex data sources that classify topological spaces. Although algebraic topology primarily uses algebra to study topological problems, application to other problems is sometimes also possible. Algebraic topology, e.g., allows for a modeling of cohort studies, i.e., clinical trials. Cohort studies are a convenient method attempting to understand correlations between particular attributes of a subject and an outcome. A good example is to determine whether the number and spatial distribution of certain immune cells and their proximity to each other or tumor cells are associated with the response to immune stimulating agents and whether this can be predicted directly from tissue specimen analysis.

The challenge posed by a vast number of practicing anatomical and surgical pathologists is increasing with a high demand for complex diagnosis using various expensive techniques. In addition, physicians and especially oncologists on behalf of their patients and the diagnostic industry are demanding the retrieval of more information from less material and the integration of all available data from many different sources to guide the treatment as part of the routine clinical approach. Furthermore, all relevant information and data are increasingly required to be available in real-time, at any time, and at every remote place. This also calls for the application of relevant related information technologies or alternative (cloud) solutions to protect data integration and patient data privacy.

The field of AI and machine-learning algorithms offers a wide array of diagnostic and therapeutic decision support. One of the prerequisite technologies is directed toward scanning a whole microscopic slide (whole slide imaging = WSI) into a high-quality digital image and applying artificial and machine intelligence-driven algorithms or solutions for further analysis [2]. This might provide a precise and reproducible anatomic assessment, leading to a robust diagnosis and a sustainable prediction of the best treatment option with a high degree of accuracy and confidence.

The development and application of sustainable image analysis solutions, algorithms, and diagnostic decision trees heavily depend on established pre- and post-analytical quality control (QC) processes that include the sufficient and standardized preparation of the specimens, the absence of any technical artifacts, continuous staining quality, etc. Each technology that provides data into a common pool of data is also a source of error, missing interfaces, unreadable formats, and different levels of robustness [3]. While genomic data are usually quite reliable and clean, transcriptomic and translational pathway readout data are more influenced by environmental and outside factors, which albeit represent part of the natural heterogeneity in tissues. Any technically prone error or missed QC will weaken or falsify the results and misguide diagnostic decisions.

Along this path the successful integration of CP and AI into the pathology information system and its connectivity with tissue diagnostic tools or platforms will start leveraging opportunities in the practice of pathology. Such a scenario allows pathology labs to go fully digital, establish attractive new business models, implement guidelines to develop standardized workflow, and promote best pathology practices. This will feed many AI and ML projects to generate decision support solutions [4]. Digital Pathology is enabling pathology to meet the demands of modern medicine and is tightly connecting pathologists with the rest of healthcare of which some areas have already undergone a digital transformation. Digital pathology is contributing to emerging fields in medicine such as immuno-oncology and the use of combination therapies. It was intended to beckon pathologists to add these new tools to their toolbox. In so doing, digital pathology will help catapult them into a future where every patient can gain access to an expert pathology opinion and all pathologists can count on computer-aided diagnostic tools [5].

Figure 1 gives a structured overview of the content of the different book chapters and the stepwise integration of AI into the pathology workflow. While sample management, assay standardization, and high-resolution image analysis are essential prerequisites to implementing AI and leveraging its full potential for today's clinical practice, the other domains allow a glimpse into the future or are yet tools in a research and developmental environment.

One of the most important but equally undervalued principles for the successful implementation of AI into the pathology workflow is a standardized sample management. It is of pivotal importance to understand the consequences of sustainable reproducibility and robust quality procedures. Consequently, a number of AI-based image analysis tools deal with the detection of poor tissue quality, extended to the identification and exclusion of blurred regions and tissue folds.

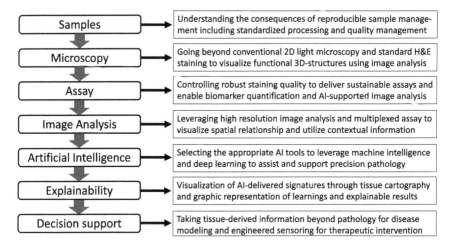

Figure 1:   Structured overview on how the following chapters will introduce AI into the entire anatomical pathology workflow and its digital solutions.

While most of the current AI-applications are still based on (ultra) thin-sliced sections of paraffin embedded tissue to allow conventional H&E stains and IHC, the imminent future of microscopy will enable the analysis of the three-dimensional structures which require advanced image analysis and computational aids to model the complexity of the tissue. Even if three-dimensional and automated imaging is the future of diagnostic microscopy, controlling a consistent and robust staining quality is necessary to deliver sustainable assays and enable biomarker quantification also through AI-supported image analysis.

With the increasing understanding of the biology and complexity of the tumor microenvironment and its heterogeneity, high-resolution image analysis and multiplexed assays allow the visualization of relevant spatial relationships and leverage the intrinsic contextual information.

But the implementing of the appropriate AI tools like deep learning and leveraging machine intelligence in general will assist and support precision pathology at all levels. However, pathologists are trained to use their visual perception and recognition of familiar patterns.

The quantification of novel and spatially oriented multi-dimensional patterns of predictive biomarkers require innovative approaches to visualize and understand AI-delivered signatures through tissue cartography and graphic representation of learnings and explainable results. The possible interpretation of "coded information" by a human mind is necessary to allow a plausibility check and the clinical adoption of digital pathology including AI-assisted decision support.

In the end, findings from individual tumors and patient cohorts may translate into nanotechnology approaches of the future. Tissue-derived information can be taken beyond pathology for disease modeling and engineered sensing for therapeutic intervention.

In summary, the key role for pathologists in tissue diagnostics will prevail and even expand through interdisciplinary work and the intuitive use of an advanced and interoperating (AI-supported) pathology workflow delivering novel and complex features that will serve the understanding of individual diseases and of course the patient.

References

[1] Fuchs, T. and Buhmann, J. (2011). Computational pathology: Challenges and promises for tissue analysis. *Comput. Med. Imaging Graph.* 35(7–8): 515–530.

[2] Campbell, W. S., Foster, K. W., and Hinrichs, S. H. (2013). Application of whole slide image markup and annotation for pathologist knowledge capture. *J. Pathol. Inform.* 4: 2, doi: 10.4103/2153-3539.107953.

[3] Gurcan, M. and Boucheron, L. (2009). Histopathological image analysis: A review. *IEEE Rev. Biomed. Eng.* 2: 147–171.

[4] Binnig, G., Huss, R., and Schmidt, G. (2018). *Tissue Phenomics: Profiling Cancer Patients for Treatment Decisions*. 1st edition, PanStanford Publishers Pte. Ltd., Singapore.

[5] Huss, R. and Coupland, S. E. (2020) Software-assisted decision support in digital histopathology. *J. Pathol.* 250(5): 685–692, doi: 10.1002/path.5388.

Chapter 2

Standardized Tissue Sampling for Automated Analysis and Global Trial Success

Eike von Leitner[*], Philipp Layer[†], and Hartmut Juhl[‡]

Indivumed GmbH, Hamburg, Germany

[*]*Von-leitner.eike@indivumed.com*
[†]*Layer.philip@indivumed.com*
[‡]*Juhl.hartmut@indivumed.com*

1. Introduction

Our ability to create, process, and analyze large datasets is revolutionizing our world. Artificial intelligence (AI) and machine learning (ML) have become powerful tools to extract information of interest from data. This results in new and innovative discoveries in all areas of our life. Naturally, it also affects medical research and medical care to some degree already. Tomorrow's medicine will be driven by the power of digitization, which becomes most visible in the field of precision medicine. Precision medicine is understood to mean an optimally tailored therapy for the individual patient in terms of the expected efficacy and the toxicity and quality of life in line with the benefit. The exponentially growing ability to recognize individual, disease-relevant factors by analyzing complex biological datasets, to put them into context with other data resources such as disease models and drug libraries, and, finally, to include all this in a diagnostic approach is breathtaking.

To utilize this tremendous opportunity for drug development and patient care, the data resources and their quality become a cornerstone for success, because AI only searches for relationships between data points and ignores whether data are right or wrong. When a data resource does not reflect the biological reality, in particular when a systematic error occurs, AI will still create results, however, they are of little value or may even harm patients. To avoid poor outcome and discreditation of digitization for patient treatment, this becomes extremely important. Because biopsies or surgically removed tissues contain the raw data for digital medicine, it is the responsibility of human pathologists to guarantee that tissue can be used for AI-driven applications in a diagnostic process that brings the full benefit of digitization and precision medicine to patients.

In oncology, precision medicine takes the highest priority, as it can lead to a fundamental change in the currently particularly unsatisfactory medical care. The current possibilities of successfully adapting oncological therapy to individual factors are miserable: in about 75% of patients, drug-based oncological treatment is unsuccessful and at the same time associated with severe side effects. As this is the disease most frequently leading to death in the Western world, the need for action is greatest here and relates both to the development of active agents and their targeted application by improving diagnostics. Due to the urgency and thus the most rapidly developing application and exploitation of opportunities, this chapter focuses on oncology. However, the approach and methodologies can in principle be applied throughout medicine.

2. AI-Advanced Data Analytics

"AI" is one of the most powerful recent inventions in data science. AI has offered numerous technologies to analyze large datasets and has now also become a meaningful tool to fight cancer. The sub-field of AI that is most relevant to large cancer datasets is machine learning, or ML. ML's purpose is to transform data and observations into knowledge and evidence-based medical decisions. Particularly, ML predictive modeling enables the prediction of an outcome of interest (e.g., type of disease, disease status, subtype of cancer, survival time, response to therapy) on tissues and patients, respectively. In addition, feature selection filters out irrelevant as well as redundant markers to identify the set of markers (called a signature) that is required for optimal prediction. While differential expression analysis examines markers in isolation, ML examines them in combination. It can sort out through hundreds of thousands of markers to find a signature and a possibly non-linear predictive model. The model can be used to inform clinical decisions (e.g., change therapy if it is predicted to be ineffective), the markers in the signature can be used to design diagnostic assays, point to plausible drug targets, and provide intuition on the underlying biological mechanisms involved.

3. Cancer — A Highly Complex, Individual Disease

In terms of genesis and variance in disease progression, "cancer" is probably the most complex disease. At the same time, recent developments in computer technology and data analytics offer a special opportunity to tackle this disease.

Over the past 50 years, the cell biological basis of cancer has been largely deciphered. We know that somatic mutations are the starting point to turn cells into cancer cells, and we now know numerous proteins, signaling pathways, etc., that are responsible for the destructive growth and ultimately lethal metastasis typical for cancer cells [1]. Numerous mechanisms have also been identified for how the body gets rid of degenerate cells and thus prevents cancer at an early stage. And it has also been possible to develop drugs against cancer cell-specific alterations, which have then been shown to be effective when used in patients. Herceptin against the HER-2 receptor in breast cancer was the conceptual proof-of-concept that scientific knowledge of cancer cell biology leads to successful precision medicine. Over the past 15 years, the development of precision medicine has been followed by the identification of various somatic mutations (K-RAS, ALK, etc.) that lead to altered cancer cell behavior that can be used therapeutically. Based on the mutation status, signaling pathways have been found whose blockade has resulted in therapeutic benefit for a subset of patients [2].

While this development is encouraging, it still reflects an analogue medicine based on the determination of one, or at least only a few, diagnostic markers, rather than attempting to map individual biological complexity through the combined evaluation of numerous data points.

Yet, this is precisely what will be needed to advance precision medicine in oncology for true patient benefit.

While genetic modifications are the key drivers of the development of cancer, and the analysis of genetic aberrations have become part of the diagnostic process, it has become obvious that in cancer (and thus its treatment), numerous other biological data points must be considered in addition to cancer genetics, such as epigenetic factors, protein expressions, and cell–cell interactions, in order to comprehensively understand the individual disease and thus target active agents (which are initially developed based on these data).

This incredible complexity, which is further greatly increased by the heterogeneous cell composition of tumors in constant flux, presents challenges for precision medicine. With today's highly sensitive and relatively inexpensive measurement techniques, combined with the opportunities brought by our era of digitization and data analysis, the prospects for success in overcoming these challenges are good.

4. Digitization of Cancer Data Quality Is Crucial

Understanding the complexity of cancer requires two types of datasets:

- The cell biological data of the individual cancer, with the help of which the biology of the tumor disease can be deciphered in order to, on the one hand, develop new specifically attacking active agents and, on the other hand, to apply these in a targeted manner only to patients in whom the cell biological prerequisites for efficacy are given.
- The clinical data of the patient (age, sex, previous diseases, concomitant diseases, family diseases, efficacy/toxicity of therapies, etc.), which help to understand the cell biological data in the individual context, which is particularly important for the development of active agents, but can also provide important indications of efficacy or toxicity before their subsequent application.

The quality of these data is critical for successful precision medicine, both for drug development and in subsequent application. In the former case, erroneous data can lead to costly drug misdevelopment, and for proper application, reproducibility of data analytics is critical [3].

Being able to collect cell biology data comprehensively and in a clinically accurate manner is arguably the greatest challenge in the advancement of precision medicine, as it requires a change in established routines in everyday clinical practice. For 150 years, oncology treatment has been based on formalin-fixed tissue embedded in paraffin (FFPE tissue), which is then evaluated histologically by the pathologist through microscopic analysis. This tissue type was also seemingly sufficient for the first steps of precision medicine, as it allows both the (immunohistochemical) detection of specific therapeutic targets (e.g., HER-2 receptor staining for Herceptin delivery) and a determination of somatic cancer-relevant mutations.

However, complex analytical technologies that determine numerous factors and calculate signatures by algorithmic approaches (e.g., multiplex staining of tissue combined with digital imaging analysis), require tissue of identical quality regarding collection and processing procedures to obtain comparable and reproducible results (Figure 1) [5].

And technologies such as proteomics and phosphoproteomics analyses necessarily require frozen, not FFPE tissue, which our healthcare system is barely prepared for.

But what is even more essential to data quality that will bring precision medicine into the digital age is how diagnostic tissue is obtained. Once the tumor is removed from the body by the surgeon, the cell biological composition undergoes significant changes due to deprivation of blood (oxygen) flow and temperature changes. Within 30 minutes of ischemia time, approximately 20% of RNA, numerous proteins, and most of the signaling molecules are present in a greatly altered quantity (Figure 2).

Stress regulatory signaling pathways are turned on or off, but proteins already used diagnostically today (e.g., EGFR and HER-2) are also subject to quantitative changes. Finally, cancer-relevant signaling pathways cannot be analyzed at all [4].

Considering how slowly formalin diffuses into tissues (1–2 mm/h), the dramatic nature of ischemia becomes clear, especially with conventional tissue preparation using formalin fixation. Thus, to base precision medicine on solid data, it is imperative that tumor tissue is immediately snap-frozen within a few minutes after surgical removal. And any immunohistochemical staining of targets or biomarker in FFPE tissue, first, requires the analysis of the protein of interest and if it remains stable under ischemia and non-standardized postsurgical tissue processing.

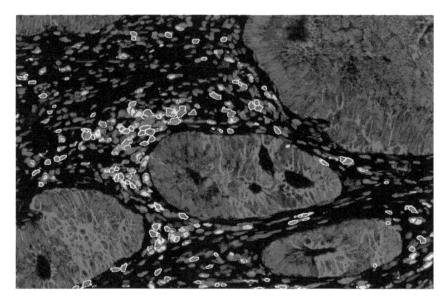

Figure 1:   The potential of digital pathology in immuno-oncology shows a CD3/CD8/ FoxP3-immune cell detection in colorectal carcinoma. PanCK staining enables subdivision of tumors into an intraepithelial area and a tumoral stroma area. An alignment of therapeutic target staining can be added (not shown). The potential read-out covers intensity of signal, distribution of cells, and spatial analysis, e.g., distance measurements between tumor and immune cells. Additional staining of specific cancer targets may also be included. Application of AI/ML-technologies allows to define prognostic and predictive signatures. However, it becomes obvious that reproducible and, thus, clinically reliable data can only be generated when the tissue quality is high. Otherwise, data between patients are not comparable and a diagnostic test would not be reproducible.

The standardized processing of tissue taking these criteria into account is the basic prerequisite for successful precision medicine, which can bring today's technical potential to full development. In the current diagnostic processes (and reimbursements), these elementary criteria are not reflected.

5. Criteria of Standardization to Enable AI-Supported Digital Medicine in Pathology

In early drug discovery stages, as well as in clinical trials, researchers are well aware that minimizing and controlling variables introduced in their

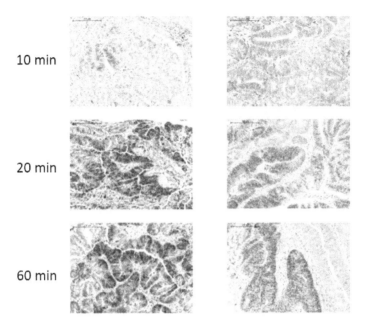

Figure 2: Phosphorylation of Erk 1/2: Different phosphorylation of key proteins of intracellular signal transduction after different times of cold post-operative ischemia.

analyses is crucial. At the same time, they are often unaware of potential confounding factors that were brought in before their experiments even started by insufficiently controlled collection of the tissue samples they analyze. But as numerous pre-analytical variables during this collection procedure itself have a major impact on the quality of the resulting data, understanding and controlling factors is crucial to producing reliable results. When it comes to imaging applications in pathology, the key factors that need to be controlled even before the "typical" aspects that are often taken into account are taking place (like staining and image acquisition) are the tissue collection itself, followed by fixation of the tissue samples.

In the past the analysis and interpretation of histopathological images solely relied on pathologists, who have been trained to detect and ignore artifacts and variability in tissue stainings. This versatility has allowed pathologists to handle individual tissue assessments well [6]. On the other hand, a lack of standardization emerged, which may lead to significant variability between tissue sections and therefore evaluations, especially in bigger studies where tissues of different origins are examined by several evaluators.

Nowadays, as there is a strong need for obtaining more reproducible and standardized results in large cohorts, the assessments of tissue sections are often performed in an automated manner, supported by AI applications that interpret the results. Therefore, there is an emerging need to eliminate potential confounding factors that might impede the ability to distinguish real biologic variability from background noise introduced by non-standardized tissue collection and handling procedures.

To ensure the overall quality, each step in the process of obtaining biological material and the patient data collected for this purpose must be precisely planned and executed. To perform valid measurements within a study, or even to go beyond that and to be able to compare results between different studies, all pre-analytical variables and processing conditions need to be tracked and documented carefully, because that data are valuable sources of information to ensure valid and sustainable results.

These considerations about standardization of tissue collection and processing will not only apply for successful drug development but also when AI-supported digital medicine is used in the clinical routine.

The following steps in pre-analytical standardization are important to consider:

- **Specimen collection**
  After surgical resection, the appropriate preparation of the sample needs to happen in close coordination with the surgeon and/or pathologist, depending on the approach of the clinical site. The localization of the tumor must be determined based on preliminary findings prior to any dissection. Representative tissue pieces of tumor and matched normal tissue should be instantly snap-frozen, while for FFPE tissue, if feasible, hollow organs should be opened to allow the fixative agent to access evenly. All individual parameters for each sample (e.g., ischemia time, distance to tumor margin) have to be assessed and documented.
- **Tissue processing**
  The objective has to be to fix the tissue within a set range of cold ischemia time, either in liquid nitrogen or formalin. Post-resection areas of tumor and adjacent normal tissue should be documented. If adjacent normal tissue is collected, a certain distance must be maintained between the tumor and the normal tissue collection site,

this distance might vary depending on the organ from which the sample was taken. The adjacent normal tissue sample can serve to represent the source of the tumor development.

Depending on planned analyses of the collected tissue, it might be useful to collect specimen pieces from mirrored samples for fresh frozen and formalin preservation, to allow comparisons to be performed between both samples. It is crucial that the tissue is fully immersed in formalin, the sizes of the samples should always remain similar, so that formalin infiltration rate into the tissue can be assessed accordingly. Those steps are often overlooked as potentially influential factors in histopathological assessments, although every step can introduce significant variability.

- **Tissue analysis**

  Being so widely used, *FFPE tissue* will continue to serve as an important resource for advancing precision medicine and digital analysis. To apply digital pathology, automated image analysis, and eventually AI-advanced image data analytics, it is essential that FFPE tissue processing, staining protocols, and automated analysis follow a stringent standardized and whenever possible automated approach.

To apply the full power of AI-advanced analytics, it will require *frozen tumor-normal tissue pairs* which can be used to obtain any layer of information from the same tissue such as genetic, transcriptomics, proteomics, phosphoproteomics, metabolomics, etc.

6. Outlook

Tomorrow's medicine will be driven by the power of digitization which becomes most visible in the field of precision medicine. The application of AI-advanced data analytics is extremely powerful for discovery and drug development as well as for its diagnostic application in pathology. Potential applications are in the field of classical histopathology using automated image analysis of complex tissue staining [6]. However, currently several projects are ongoing that search for digital imaging algorithms that identify the molecular profile of tumors and, subsequently, a targeted therapy by analyzing simple H&E stainings. For these applications, it is obvious that highly standardized processes are important to obtain reproducible results, which start with the tissue collection procedures as described above but also include the minimization of artifacts because of poor tissue block production or staining techniques.

The full power of AI-driven data analytics will be applied if frozen tissue is available and if this tissue is snap-frozen within minutes after surgical removal to preserve the biological and molecular reality (as described above). The possibility to integrate all biological information from a tumor tissue in comparison to the individual normal tissue is of enormous power, not only to identify new targets or develop drugs more efficiently but also for a precision medicine approach [7].

However, the outcomes of AI-advanced data analytic technologies for the understanding of biological datasets depend on datasets which reflect the reality of biology. And the tissue quality becomes thereby the most critical factor in the advancement of precision medicine.

Realistically, the required high standard of tissue collection and processing cannot be applied in the foreseeable future in clinical routine. However, it is applicable in specialized cancer centers, which are the drivers of precision medicine, and visibility of its impact on cancer care will finally also change clinical practice which, currently, is still obstructed by reimbursement guidelines — the most challenging guard to advance medicine.

References

[1] Vogelstein, B., Papadopoulos, N., Velculescu, V. E., Zhou, S., Diaz Jr., L. A., and Kinzler, K. W. (2013). Cancer genome landscape. *Science* 339: 1546–1558.

[2] Chin, L., Andersen, J., and Futreal, P. (2011). Cancer genomics: From discovery science to personalized medicine. *Nat. Med.* 17: 297–303.

[3] Compton, C. (2007). Getting to personalized cancer medicine. *Cancer* 110: 1641–1643.

[4] David, K. A., Unger, F. T., Uhlig, P., Juhl, H., Moore, H. M., Compton, C., Nashan, B., Dörner, A., de Weerth, A., and Zornig, C. (2014). Surgical procedures and postsurgical tissue processing significantly affect expression of genes and EGFR-pathway proteins in colorectal cancer tissue. *Oncotarget* 5: 11017–11028.

[5] Koelzer, V. H., Sirinukunwattana, K., Rittscher, J., and Mertz, K. D. (2019). Precision immunoprofiling by image analysis and artificial intelligence. *Virchows Arch.* 474: 511–522.

[6] Webster, J. D. and Dunstan, R. W. (January 2014). Whole-slide imaging and automated image analysis: Considerations and opportunities in the practice of pathology. *Vet. Pathol.* 51(1): 211–223, doi: 10.1177/0300985813503570.

[7] Lakiotaki, K., Georgakopoulos, G., Castanas, E., Røe, O. D., Borboudakis, G., and Tsamardinos, I. (2019). A data driven approach reveals disease similarity on a molecular level. *Syst. Biol. Appl.* 5: 39.

Chapter 3

Light-Sheet Microscopy as a Novel Tool for Virtual Histology

René Hägerling[*,†,‡,§,¶] and Fabian Mohr[*,§,‖]

Charité — Universitätsmedizin Berlin, Corporate Member of Freie Universität Berlin and Humboldt-Universität zu Berlin, Institute of Medical and Human Genetics, Augustenburger Platz 1, 13353 Berlin, Germany

†*Berlin Institute of Health at Charité — Universitätsmedizin Berlin, BIH Center for Regenerative Therapies (BCRT), Charitéplatz 1, 10117 Berlin, Germany*

‡*Berlin Institute of Health at Charité — Universitätsmedizin Berlin, BIH Academy, Clinician Scientist Program, Charitéplatz 1, 10117 Berlin, Germany*

§*Charité — Universitätsmedizin Berlin, Project Team "3D-HistoPATH", Augustenburger Platz 1, 13353 Berlin, Germany*

¶*Rene.Haegerling@charite.de*
‖*3D-HistoPATH@charite.de*

1. Introduction

The current gold standard in histopathology is based on the visualization and analysis of representative two-dimensional (2D) tissue sections of a three-dimensional (3D) specimen. It has been established for many decades and is currently used for most histological examinations. The diagnostic workflow for histology consists of three steps: sample preparation, visualization, and partly algorithm-aided analysis[1].

For sample preparation, tissue biopsies or parts of a tissue preparation are examined macroscopically and subsequently cut for transfer of relevant tissue into casting molds. The casting molds are then poured with paraffin or a resin and prepared for manual cutting of the tissue blocks. Depending on the required staining, histological tissue sections of various thicknesses (e.g., 1 up to 10 μm) are generated from the tissue blocks. As the histological sectioning process still represents a manual step in the partially automated histology workflow, tissue blocks require manual clamping into the cutting device as well as cutting to reach relevant areas of the tissue preparation. The individual tissue sections are unrolled with a fine brush and finally transferred onto a glass slide. To be able to histologically section additional representative areas of the sample and thus include them in the analysis, large parts of the preparation are discarded, before other relevant areas are accessible and transferred onto the next slide. Subsequent to the manual sectioning process, the individual sections are processed in an indication-specific manner using chemical (e.g., Hematoxylin & Eosin (H&E) staining) or immunohistochemical staining procedures using automated staining devices. After staining, representative tissue sections are sequentially analyzed by pathologists using standard or fluorescence light microscopy.

Despite new advances in the field of staining procedures, e.g., the introduction of *in situ* hybridization techniques, the classical 2D histological gold standard is still associated with a number of limitations that complicate the interpretation of histological sections and therefore complicate the diagnosis.

Physical cutting of tissue blocks inevitably results in tissue distortion as well as tissue loss, which significantly complicates subsequent analysis and interpretation. In addition, when large parts of the tissue are discarded, material is no longer available for further examinations such as staining or deoxyribonucleic acid (DNA) extraction.

Further, sections are only used for staining of individual markers. Multiple stainings of the same section are rather the exception. Hence, multiple markers are only visualized in consecutive sections, which further complicates the interpretation.

In addition to the limitations mentioned, a missing or insufficient spatial information leads to an information-based reduction of the understanding of complex pathologies. Especially in the interpretation of complex 3D structures such as the vasculature or the tumor microenvironment, a spatial understanding of tissue architecture and cellular composition is essential for an improved understanding of the underlying pathology and consequently key to diagnosis and selection of suitable next-generation therapeutic approaches.

Since the limitations are inherent to the physical sectioning method, they can only be overcome by identifying new technologies and approaches that allow studying and analyzing tissue samples in their entirety [2]. Therefore, instead of physical sectioning methods, optical sectioning techniques provide an elegant alternative. However, ultrasound, magnetic resonance imaging (MRI), optical projection tomography, or (computed tomography) CT-based techniques do not provide a cellular resolution combined with a high-magnification, which is essential for an application in histology.

Current technology in laser microscopic methods such as confocal or multiphoton microscopy are only insufficient alternatives. Although both methods provide high-resolution imaging of individual optical sections and in the X–Y plane, there is a blurring in the Z plane due to the optical limitation of the method, i.e., the excitation and detection of fluorescent signals pass through the same objective. Since the axial resolution is many times lower than the lateral resolution, this leads to anisotropy and the associated lower quality of the 3D reconstruction of the image data. To illustrate this in other words: The representation of a perfectly spherical body would result in high-resolution round circles at the level of 2D optical sections. However, a subsequent 3D reconstruction of the optical sections leads to an ellipsoidal representation of the spherical body. Although new approaches of image deconvolution try to address this problem, a sufficient 3D representation is still not possible. Therefore, large-volume structures can only be represented inadequately in 3D space by using these methods (Figure 1(A)). Confocal and multiphoton microscopy are point scanners which only detect individual pixels instead of simultaneously recording an entire optical plane. This results in reduced imaging speed and therefore they are limited in their use for 3D tissue samples in routine histopathology.

However, mesoscopic imaging technologies which unify advantages of high-resolution laser scanning microscopy and macroscopic imaging

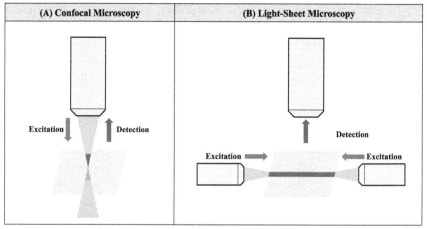

Figure 1:   Comparison of Confocal and Light-Sheet Microscopy. A schematic representation of excitation and detection pathways in Confocal (A) and Light-Sheet Imaging (B) as well as relevant parameters. Direction of excitation and emission pathways are marked by blue and red arrows.

techniques are emerging as novel approaches in 3D-histology [2, 3] (Figure 1(B)).

2. Technology Description

Light-Sheet Microscopy is a mesoscopic imaging modality which is suitable for optical sectioning of human tissue samples. This microscope creates a series of optical sections covering the entire tissue sample (for representative single optical section, see Figure 2). Depending on the required resolution, the step size between each optical section can range from less than one micron to several microns. By performing digital 3D reconstruction of the captured optical sections (up to several ten thousands), the entire tissue sample is visualized in 3D space with cellular resolution [2–5].

Since the excitation by the light-sheet and the detection of the signal is achieved by different, typically perpendicularly aligned objectives, the

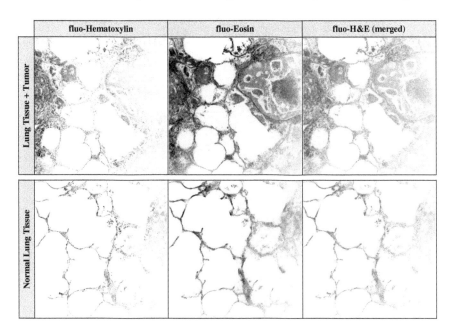

Figure 2:   Single optical sections and virtual fluo-Hematoxylin & Eosin (fluo-H&E) staining of human lung tissue biopsies. Single representative optical sections of wholemount stained lung biopsy visualized using a Light-Sheet Microscope. Normal lung tissue and lung tissue including tumor tissue optical sections were stained using fluorescent Hematoxylin & Eosin analogs. Single channels are shown as merged virtual fluo-H&E stainings.

limitations described for confocal and multiphoton microscopy do not occur in the 3D reconstruction of large tissue samples. Thus, an isotropic to nearly isotropic resolution can be achieved, i.e., the lateral and axial resolutions are the same or similar. Since this is a crucial parameter for the 3D reconstruction of large tissue samples, entire samples of several cubic centimeters in size can be visualized and digitally reconstructed in a 3D format. This procedure has already been successfully described for 3D reconstructions of samples from animal models as well as human tissues [2, 6, 7]. A representative 3D reconstructed sample can be found in Figure 3.

    To perform a Light-Sheet Microscopy-based analysis, several prerequisites for staining and analysis of a sample are necessary.

    For the identification of relevant structures and biomarkers within a sample, chemical staining (e.g., H&E) as well as IgG antibody-mediated

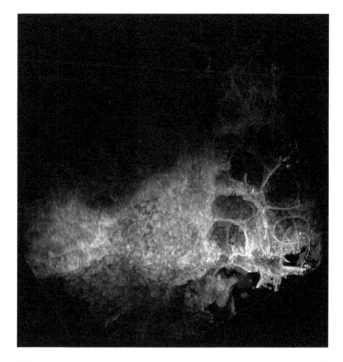

Figure 3:   3D reconstruction of pulmonary metastasis of colorectal cancer. A fluorescent Hematoxylin & Eosin-stained human lung biopsy was visualized using Light-Sheet Microscopy. Shown is a digital 3D reconstruction of the entire biopsy highlighting the tumor (red arrow) within the biopsy.

immunofluorescence staining are necessary. In contrast to chemical dyes, which can fully penetrate the entire sample within seconds to minutes, IgG antibody stainings require longer incubation times, which can vary from a few days to weeks depending on tissue type and sample volume. This limits the current use of the technology, since only a few of the antibodies established in pathology are available for wholemount staining. Novel antibody stains and chemical analogs to standard histology stains must be developed to cover the wide spectrum of currently available markers.

By combining different secondary dyes, multiplexing can be performed to stain several biomarkers and structures and to be imaged in the same sample. Currently, this is only limited by the number of available fluorescent secondary dyes, emission filters, and excitation wavelengths within the Light-Sheet Microscope.

The most crucial requirement for illumination in Light-Sheet Microscopy is based on the optical transparency of the tissue. Since the

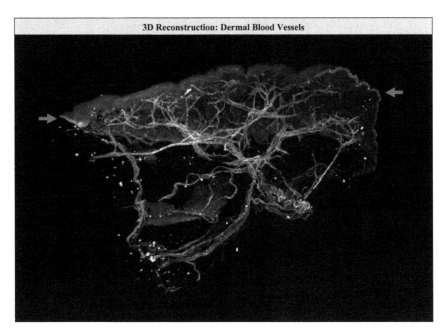

Figure 4:   3D reconstruction of dermal blood vasculature in human control samples. Optical projection of a digital 3D reconstruction of the human blood vasculature in a healthy skin biopsy. Red arrows indicate epidermis.

light-sheet can only deeply penetrate optically transparent structures, tissue samples must be optically cleared before analysis to produce a high-quality image. Optical clearing is defined as the reduction of optical inhomogeneities within a sample and the associated scattering of light by matching the refractive index (RI) of the tissue to the surrounding medium (RI matching). Since biological samples have a high refractive index (RI ~ 1.50), organic solvents with a similar RI are suitable. Since endogenous fluorescence of fluorescent reporter proteins (which are often used in experimental models) does not need to be obtained for human samples, optical clearing with a benzyl alcohol–benzyl benzoate (BABB, RI ~ 1.56) solution is an ideal clearing medium for these samples [8]. In addition, BABB-based clearing methods are also compatible with chemical and immunofluorescence staining. To achieve the best possible recording quality, complete optical transparency is necessary. A perfect optically cleared sample with stained blood vasculature is shown in Figure 4.

Since some pigments or light-absorbing structures are present in tissues, passage of the light-sheet through a sample can be blocked, and hence shading, i.e., non-illuminated areas, can occur. For this reason, some Light-Sheet Microscope systems have the option of moving the light-sheet slightly during the exposure to shine past the light-blocking structure and achieve complete illumination of the optical plane.

3. Data Management and Image Analysis

Light-Sheet Microscope recordings are widely based on the standard OME.tiff [9] format, with a few exceptions for some vendors. The size of such files is massively larger compared to regular histology files and can range up to >50–100 GB per sample. Based on standard file-formats, some software tools used in z-stack analysis from confocal microscopy data can also be used on Light-Sheet Microscope derived files. However, as the high volume of data requires a stronger IT infrastructure compared to regular histology data, it remains unclear whether current digital pathology solutions or Picture Archiving and Communication Systems (PACS) used in clinical routine are suitable for handling these data files.

In Figure 5, the data associated tasks from data input to visualization and a final report are summarized and will be explained in detail in what follows.

Several quality assurance tasks are performed before using Light-Sheet Microscopy imaged samples for analysis. This includes measurements to confirm correct staining, e.g., dye distribution and linear staining pattern of antibodies, or to adjust samples accordingly beyond 2D imagery in a 3D volume. In addition, metadata management to link used dyes to their respective biological targets is required to make sense of the samples in an automated way, since fluorescent signals cannot be pinned easily to a staining, as compared to regular H&E staining in traditional histology. Further, if staining protocols are performed incorrectly or with variation, inter-sample adjustments when analyzing large cohorts of samples for comparison reasons need to be performed.

Sections of standardized 3D volumes can be visualized using conventional microscopy software for a first impression based on e.g., fluorescent H&E staining (Figure 3). For more elaborate analysis, algorithms can be used based on marked regions of interest (RoI) within a sample. In

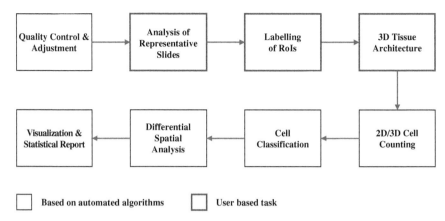

Figure 5: Data workflow for analysis of Light-Sheet Microscopy-derived imaging samples. Samples are automatically quality controlled and adjusted. Based on user input, particular regions of interest are analyzed and visualized in combination with tissue architecture. Cell counts are based on virtual slides and 3D volumes, and cells are classified using machine learning algorithms. Statistical analysis and visualization are derived from obtained data.

representative virtual 2D sections of the sample, pathologists can mark various tissue phenomena including (but not limited to) tumors, tissue structures such as blood vessels or basal membranes, or regions completely based on biomarker stainings. The virtual sections can be analyzed and labeled in a traditional X–Y view but also in the different X–Z and Y–Z angles.

Based on 3D reconstruction algorithms and manual labeling the tissue architecture is computed and rendered into a voxel (represents the respective intensity value on a 3D regular grid) octree. This allows for a sensible visual detail while being tightly compressed in spaces which are lowly populated [10]. Independent from tissue architecture, algorithms are used to identify cells based on virtual 2D sections, like already existing algorithms used in conventional digital pathology. Combined with algorithms segmenting the entire volume data, the likelihood of correct identification and count of cells is increased. Cells are further classified according to visual morphology and stained markers based on volume data. Ultimately, a spatial analysis of cell distribution within the tissue architecture is performed resulting in data of which and how many cells reside within different compartments of the tissue, e.g., CD8$^+$ (labeled) T-cells within a tumor and in stroma.

Based on the derived data, 3D visualization can be performed resulting in adjustable models including representation of various architecture components and cell types. Heatmaps can be generated to illustrate model accuracy and tissue distribution of cell types. Further, statistical analysis based on cell position, counts, and cell types can be performed.

4. Discussion and Summary

Light-Sheet Microscopy represents a novel and groundbreaking tool for future diagnostics based on histopathology. The technology has the potential to overcome the limitations associated with physical sectioning of tissue samples including tissue distortion, manual slicing and transfer to glass slides, tissue loss, and spatial resolution. Especially in samples with low material availability, Light-Sheet Microscope-based imaging allows for down-stream applications, e.g., DNA extraction or regular 2D histology.

By performing optical sectioning of previously fluorescence-stained and optically cleared human tissue samples, visualization of not only single sections but of entire specimen of several cubic centimeters in size with cellular resolution is possible. By combining fluorescent chemical dyes (e.g., fluo-H&E) and fluorescence-labeled IgG antibodies, tissue samples can be stained for various biomarkers and cellular compartments, and visualized simultaneously or sequentially using the Light-Sheet Microscope. Multiplex analysis, which is only limited by the number of lasers and filters in the microscope, allows detailed characterization of the tissue samples in a routine setting.

As the analysis is not only limited to optical sections in the X–Y axis but allows vision also in the X–Z and Y–Z axis, it enables pathologist to analyze samples from various perspectives. In addition, the digital 3D reconstruction of the entire specimen allows inspection and interpretation of anatomical structures and cells in their spatial context. This will allow 3D determination of treatment-relevant parameters, such as the distance between tumor and resection margins for tumor surgery or identification of individual tumor cells or micro-metastasis in biopsies.

Based on volumetric data, AI-algorithms will be more sensitive and specific. Additional algorithms will enable the read-out of digital biomarkers in a spatial context leading to better patient identification for treatment for e.g., precision oncology.

In the future, more staining protocols, similar to fluo-H&E, will be developed to resemble current histology stains and increase the spectrum of analysis based on Light-Sheet Imaging. As staining protocols, clearing methodology, and computerized analysis on virtual slides are already possible in 2D-based histology, a fully automated process from sample entry to results and a final diagnosis based on Light-Sheet Microscopy seems feasible. The histopathology process will be completely automated based on Light-Sheet Microscopy, since glass slides and hence physical sample slices are obsolete.

Acknowledgments

We thank Prof. Bruno Märkl for providing tissue samples and reagents as well as the fruitful discussions toward establishing the assays.

References

[1] Niazi, M. K. K., Parwani, A. V., and Gurcan, M. N. (2019). Digital pathology and articifical intelligence. *Lancet Oncol.* 20(5): e253–e261.

[2] Hägerling, R., Drees, D., Scherzinger, A., Pollmann, C., Martin-Almedina, S., Butz, S., Gordon, K., Schäfers, M., Hinrichs, K., Ostergaard, P., Vestweber, D., Görge, T., Mansour, S., Jiang, X., Mortimer, P. S., and Kiefer, F. (2017). VIPAR, A quantitative approach to 3D-histopathology applied to lymphatic malformations. *JCI Insight* 2(16): e93424.

[3] Huisken, J. and Stainier, D. Y. R. (2009). Selective plane illumination microscopy techniques in developmental biology. *Development* 136(12): 1963–1975.

[4] Mertz, J. (2011). Optical sectioning microscopy with planar or structured illumination. *Nat. Methods* 8: 811–819.

[5] Keller, P. J. and Dodt, H. U. (2012). Light-sheet microscopy of living or cleared specimens. *Curr. Opin. Neurobiol.* 22(1): 138–143.

[6] Feuchtinger, A., Walch, A., and Dobosz, M. (2016). Deep tissue imaging: A review from a preclinical cancer research perspective. *Histochem. Cell Biol.* 146(6): 781–806.

[7] Hägerling, R., Pollmann, C., Andreas, M., Schmidt, C., Nurmi, H., Adams, R. H., Alitalo, K., Andresen, V., Schulte-Merker, S., and Kiefer, F. (2013). A novel multistep mechanism for initial lymphangiogenesis in mouse embryos based on ultramicroscopy. *EMBO J.* 32: 629–644.

[8] Orlich, M. and Kiefer, F. (2018). A qualitative comparison of ten tissue clearing techniques. *Histol. Histopathol.* 33(2): 181–199

[9] Goldberg, I. G., Allan, C., Burel, J. M., Creager, D., Falconi, A., Hochheiser, H., Johnston, J., Mellen, J., Sorger, P. K., and Swedlowet, J. R. (2005). The open microscopy environment (OME) data model and XML file: Open tools for informatics and quantitative analysis in biological imaging. *Genome Biol.* 6(5): R47.

[10] Jackins, C. L. and Tanimoto, S. L. (1980). Oct-trees and their use in representing three-dimensional objects. *Comput. Graphics Image Process* 14(3): 249–270.

Chapter 4

Stain Quality Management and Biomarker Analysis

Søren Nielsen

NordiQC, Institute of Pathology, Aalborg University Hospital,
Ladegaardsgade 3, DK-9000 Aalborg, Denmark

sn@rn.dk

1. Introduction

Immunohistochemistry (IHC) is a well-established and indispensable assay being consistently performed in anatomic pathology worldwide in order to provide a specific diagnosis and subclassification of neoplasms [1, 2]. IHC serves at present both as a diagnostic, prognostic, and predictive assay and the results contribute to the final decision for patient treatment. IHC is in its most common application a descriptive, threshold-based test to determine if a target analyte/protein is present or absent and thus the result, classified as either positive or negative, is used in the final diagnostic work [3, 4]. IHC is widely applied in formalin-fixed paraffin embedded (FFPE) histological material and primarily used to identify the origin and subtype of cancers in, e.g., solid tumor diagnosis and hemato-pathology. At present, more than 200 IHC targets such as Thyroid Transcription Factor 1 (TTF1), Cytokeratins (CKs), Cluster of differentiation (CD) molecules, etc., are of central importance for cancer classification and consequently IHC assays for these targets must be implemented in virtually all pathology laboratories in their testing repertoire. In the era of precision medicine, IHC is increasingly applied and as such transitioning to be a biomarker method [4]. The terminology "biomarker" is broadly defined as any biological or physiological test that is used to identify disease, guide selection of targeted therapy, or monitor disease reoccurrence [5]. In the last decades, the number of predictive IHC tests have gained increasing acceptance and been widely applied for decision-making prior to the administration of targeted therapies. Since the introduction of HercepTest™ (Dako) in 1998 as an IHC assay for the demonstration of human epidermal growth factor receptor 2 (HER-2) protein overexpression in breast cancer and thereby guide the selection of patients to be treated with Herceptin® (Roche/GenenTech), many similar IHC assays have been developed and now used as Companion Diagnostic (CDx) assays to stratify patients and predict the response likelihood of a specific drug [6]. Examples of these include IHC analysis of anaplastic lymphoma kinase (ALK) and ROS1 mutations in non-small cell lung cancer (NSCLC) for treatment with tyrosine kinase inhibitors as Crizotinib or Alectinib and more recently introduction of IHC assays to demonstrate the level of PD-L1 in many cancer types such as NSCLC, urothelial carcinoma, triple negative breast carcinoma (TNBC) to predict response with checkpoint inhibitors such as Pembrolizumab, Azetolizumab, and comparable drugs [7, 8].

IHC is in general a descriptive and qualitative test generating results to be reported as negative or positive by the investigator using an identified and accepted threshold to separate the two entities. However, for some IHC tests and particularly for predictive IHC assays a more complex readout must be performed and by using predefined and validated cut-off values and scoring criteria, e.g., based on the percentage of cells being positive, localization in the cell, and type of cells, the IHC result can be reported as positive or negative.

It is indisputable that biomarker testing by IHC is a critical element in patient care in the era of precision medicine, and in general IHC is a robust, reproducible, and reliable method. Nevertheless, IHC is a complex method, where the end result is influenced by multiple parameters in the pre-analytic phase, the analytic phase, and the post-analytic phase and the method applied by the laboratory for each of the multiple steps in the three phases [9] (Table 1). At least 4 million different protocols can be generated for an IHC analysis of one specific target analyte inducing a great risk of low inter- and intra-laboratory reproducibility for the IHC analysis [10].

Table 1:   The central multiple parameters influencing the IHC result.

Pre-analytical phase	Analytical phase	Post-analytical phase
Sample acquisition	Epitope retrieval	Scoring/readout
• Method for biopsy	• HIER, time, buffer, pH	• Cut-offs, cell of focus
• Transportation	• Proteolysis, time, enzyme	• Pathologist, digital analysis
Fixation	Antibody	Controls
• Ischemia	• Monoclonal, polyclonal	• Negative, positive controls
• Type and time	• Clone, titre, buffer	• Type; tissue, cell lines
• Ratio to tissue	Detection	• On-slide, batch
Decalcification	• Sensitivity, specificity	
• Type and time	Chromogen	
Tissue	• Localization, enhancement	
• Block age, storage	Instrumentation	
• Section thickness, storage	• Manual, automation (type)	

From this perspective, central challenges for IHC exist and it can with good reason be asked if it is possible to standardize IHC among diagnostic and clinical IHC laboratories to produce accurate and precise results. At this point that should not be the goal, as it is impossible to standardize all IHC methods with the objective of only using one universal methodology. The objective should rather be providing guidance to laboratories and also IHC assay vendors to apply best practices for IHC protocol development, optimization, and validation in the journey and hereby transition of an antibody to an IHC biomarker with harmonization of end results. In addition, implementation and use of an IHC quality management program for the total IHC test ranging from internal quality control, participation in external quality programs, and a quality maintenance/improvement system is highly recommended and required within the individual laboratories performing IHC for "precision testing" in cancer diagnostics [11, 12].

This chapter will highlight the most important requirements of internal quality control and experiences observed from external quality programs for the IHC analytical phase and also address the critical elements of the post-analytical phase. The requirements addressed are typically based on proposed published recommendations and guidelines from regulatory bodies, expert committees, interest groups, and organizations involved in pathology practice, e.g., American Society of Clinical Oncology (ASCO), College of American Pathologists (CAP), Clinical and Laboratory Standards Institute (CLSI), International Association for the Study of Lung Cancer (IASLC), International Society for Immunohistochemistry and Molecular Morphology (ISIMM), etc.

2. Internal Quality Control

2.1. *Purpose and use of IHC*

In 2000, Clive Taylor (the father of IHC) published an editorial review with the title "The Total Test Approach to Standardization of Immunohistochemistry" in which he emphasized *IHC is technically complex, and no aspect of this complexity can be ignored, from the moment of collecting the specimen to issuance of the final report* [9]. Even now, after more than 20 years of its publication, this statement is the backbone of internal quality control and should be integrated in the entire process to implement an IHC biomarker for diagnostic use.

For each IHC biomarker to be implemented, attention and considerations to intended use, diagnostic field(s), and scoring criteria for defining a positive vs. negative result must be addressed. The intended use, or in other words, the purpose of the IHC test, is the core element of the development, calibration, and validation of a new IHC assay and must be identified prior to the initiation of the validation process. If the use and purpose(s) are not aligned in the developmental and validation process or are changed after validation, this might induce a risk of inaccurate IHC results. An evident example of the importance of aligning use and purpose has been observed for IHC for Anaplastic lymphoma kinase (ALK) as this initially was developed by laboratories and vendors of IHC assays with the intended use to identify anaplastic large cell lymphoma with ALK gene rearrangements but subsequently also a need and request to detect ALK expression in NSCLC for personalized medicine with tyrosine kinase inhibitors was established [13, 14]. However, the two cancer types possess different gene rearrangements with different protein expression levels and it was discovered that the initial ALK IHC methods validated for ALK expression in lymphoma were unreliable and provided aberrant false-negative results in NSCLCs harboring reduced levels of the ALK fusion protein and consequently required an IHC protocol with increased analytical sensitivity for the new purpose [15].

The importance of alignment of use and purpose of IHC biomarker testing has in particular been emphasized for predictive PD-L1 IHC assays in immunotherapy. In the era of precision medicine, the one-drug and one-diagnostic assay strategies have paved the way for approval of therapeutic drugs and are the cornerstone of CDx assays. For PD-L1 IHC, several CDx assays are available to determine patient eligibility for a specific targeted therapy. Each PD-L1 CDx assay has its own characteristics and requirements for accessing specific IHC instrumentation, IHC methodology, staining patterns, and scoring criteria, which complicate and challenge the use in a diagnostic pathology laboratory. Several studies and meta-analyses have been performed to investigate the possibility of interchangeability of the CDx assays for PD-L1 IHC and indicate that despite some CDx assays showing similar analytical concordance in certain cancer types, there are data clearly indicating a risk of misclassification on the PD-L1 status and thereby inappropriate patient treatment strategy if used in other cancer types with both similar or altered readout criteria for the different cancer types [16–19]. This underlines the need for the laboratory performing IHC to always align use and purpose of IHC biomarker testing to secure accurate results for diagnostics.

2.2. *The tissue toolbox*

Following the identification of the purpose of an IHC test, the journey from an antibody to an IHC biomarker test is based on a series of steps from protocol development, optimization, validation, and establishment of monitoring tools for the reproducibility of the IHC test when implemented for routine diagnostics [20, 21].

There is no widely applied and standardized approach as the road map for laboratories to develop and validate IHC biomarkers for diagnostics, and the specific requirements to meet quality standards for validation of IHC biomarkers are different on national and international basis. Accordingly, by the standard ISO 9000, validation is defined as *confirmation, through the provision of objective evidence, that requirements for a specific intended use or application have been fulfilled.* The element to provide objective evidence can in IHC be provided by collecting evidence that the relevant test performance characteristics (TPCs) have been identified, assessed, and fulfilled for the IHC test in question [22, 23]. As IHC is a descriptive and qualitative test, the TPCs should provide information for the IHC test concerning (1) impact and robustness of pre-analytical parameters, (2) basic level of technical and analytical sensitivity and specificity, (3) analytical accuracy (sensitivity/specificity), (4) relevant reportable expression range, and (5) tools to evaluate reproducibility [22].

In order to meet the specifications of these five TPCs, it can be of high value to perform the process for IHC test development on tissue micro arrays (TMA) with different normal and neoplastic tissues being collected and processed by conditions similar to the diagnostic materials for which the IHC test subsequently shall be used for. To facilitate the journey from antibody to an IHC biomarker test, the construction and application of the "tissue toolbox" as illustrated in Figure 1 are instrumental and a cornerstone in the quality management of IHC biomarkers. Throughout the developmental and validation phase of IHC tests, access to and inclusion of "iCAPCs" (Immunohistochemical Critical Assay Performance Controls) plays a pivotal role [24]. The concept of iCAPCS was mainly developed by the International Ad Hoc Expert Committee for IHC and refers to normal (preferable) human tissues being well characterized concerning predictable expression levels, reaction patterns, and cellular localization for the target analyte. ICAPCS shall in essence be used to evaluate the level of analytical sensitivity with focus on limit of detection (LOD), level of technical and analytical specificity, and, finally, providing data on the reproducibility of the IHC test. Typically, 2–3 tissues can serve as iCAPCs and must include structures/cells with different expression levels ranging from high expression level — verifying correct IHC test being performed (e.g., right antibody applied); low expression level — verifying

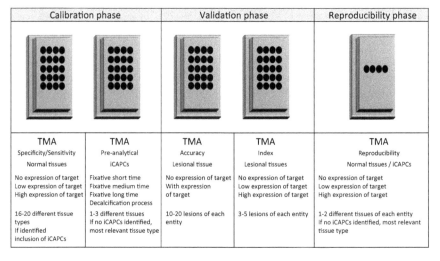

Calibration phase		Validation phase		Reproducibility phase
TMA	TMA	TMA	TMA	TMA
Specificity/Sensitivity	Pre-analytical	Accuracy	Index	Reproducibility
Normal tissues	iCAPCs	Lesional tissue	Lesional tissues	Normal tissues / iCAPCs
No expression of target	Fixative short time	No expression of target	No expression of target	No expression of target
Low expression of target	Fixative medium time	With expression	Low expression of target	Low expression of target
High expression of target	Fixative long time	of target	High expression of target	High expression of target
	Decalcification process			
16-20 different tissue types	1-3 different tissues	10-20 lesions of each entity	3-5 lesions of each entity	1-2 different tissues of each entity
If identified inclusion of iCAPCs	If no iCAPCs identified, most relevant tissue type			If no iCAPCs identified, most relevant tissue type

Figure 1:   The TMA tissue toolbox — components and objectives.

LOD being accomplished; and no expression — verifying basic technical and analytical specificity.

2.3. *Calibration phase*

Initially, all new antibodies intended to be implemented for diagnostic use in a laboratory must be subjected to a calibration phase to identify the "best practice" IHC protocol that gives a technical optimal result with focus on signal-to-noise ratio, morphology, impact of pre-analytical conditions, and providing an overall performance with a reaction pattern as expected. At this stage, it is crucial to collect information about the expected reactivity of the antibody in a broad range of clinically relevant tissues and lesions. Descriptions of staining patterns, both for the sub-cellular (nuclear, cytoplasmic, membranous) and tissue distribution are essential for the evaluation. When an antibody is tested as replacement for an existing marker, the laboratory has a set of benchmark data that facilitates the interpretation of a comparative study of a new vs. old antibody, whereas the process is much more challenging for a brand new marker with no history and benchmark data available in the laboratory and an increased number of additional considerations have to be taken into account.

Sources to gain data on expected reaction patterns, recommendations on tissue controls, and general observations for the antibody and target analyte are manifold. The first and most applicable source being PubMed, www.pubmed.ncbi.nlm.nih.gov, with access to a plethora of publications and scientific studies. In addition, a few suggestions are as follows (not prioritized); (1) vendor package inserts, (2) the human protein atlas (www.proteinatlas.org), (3) www.antibodypedia.com, and (4) www.nordiqc.org.

When the expected reaction pattern and appropriate positive and negative tissue controls have been identified and described, the antibody should be tested on a TMA with, e.g., 16–20 different normal tissues to evaluate the analytical sensitivity and specificity. If iCAPCs are available and identified, these must be included in the TMA or tested separately. At Aalborg University Hospital, Denmark, domicile of the international external quality program NordiQC, a TMA containing samples of skin, esophagus, bladder, uterine cervix, tonsil, spleen, breast, kidney, liver, pancreas, adrenal gland, thyroid gland, colon, lung, prostate, cerebrum, and fallopian tube is used as a primary tool in the calibration phase.

All tissues used must be processed accordingly to routine and standard pre-analytical conditions used by the laboratory. Serial sections of such "sensitivity/specificity TMA" are effective to evaluate the performance of different protocols for a new primary antibody and are valuable to identify the best method for the subsequent validation phase.

Optimally, in the calibration phase the IHC protocol is tested on a pre-analytical TMA with iCAPCS or other relevant normal tissues processed by the different relevant pre-analytical conditions being applicable in the laboratory in order to evaluate the robustness of the IHC protocol in these conditions. According to Engel *et al.* [25], more than 60 pre-analytical variables have been identified and can potentially affect the outcome of an IHC test. However, not all variables have been documented to impact the results and for some the data show conflicting reports [26]. Strongest agreement in publications have shown that time to fixation, fixative composition, fixation time, and decalcification process are the most critical parameters affecting IHC — in this context, also critical for molecular-based techniques as next gene sequencing (NGS), real-time polymerase chain reaction (RT-PCR), and fluorescence *in situ* hybridization (FISH).

The conditions of tissue handling from surgery to fixation include warm ischemia, e.g., vessels clamped at surgery and cold ischemia, e.g., time from removal of sample to fixation, sample temperature prior fixation, and is both related to the physical process applied and time. For IHC biomarkers, cold ischemia conditions have been documented to impact the final result, despite conflicting reports also having been published indicating no or only minimal impact and the diverging results most likely related to the different targets and tissue types included in the studies [27]. In breast pathology, publications have indicated that estrogen receptor (ER), progesterone receptor (PR) and HER-2 expression can be affected and changed due to cold ischemia [28–30]. The reported changes caused by cold ischemia are primarily characterized by reduced expression and thereby potential misclassification of a tumor, or need for additional diagnostic work-up to define the most adequate treatment strategy. This can be the scenario for a HER-2 IHC test in tissue subjected to extensive cold ischemia and the HER-2 IHC status changes from 2+ to 1+ with the tumor thereby not being investigated by *in situ* hybridization (ISH) as indicated by HER-2 IHC testing guidelines and consequently not evaluated for HER-2 positivity as defined by ISH (ASCO/CAP HER2). Similarly for ER, where a reduced ER expression caused by ischemia changes the ER result from ER positive (>10% cells positive) to ER low (1–10% cells positive) and supplemental tests indicated and needed as panel-based gene-expression assays, like Oncotype DX (Genomic Health, Redwood City, CA) to assess the likelihood of clinical benefit offered by endocrine therapy when added to chemotherapy.

Cold ischemia will not always lead to decreased expression levels for the target analyte and can also cause increased expression levels as observed for phosphorylated targets and notably alter the morphology compromising the microscopic interpretation of the lesion [31].

In order to standardize and minimize the impact of cold ischemia, organizations such as ASCO/CAP and IASLC have provided guidelines based on the published data and meta-analyses and recommend that tissue should be immersed in fixative as soon as possible and indicate an optimal range of maximum 30–60 min. from sampling till immersion in fixative.

The temperature during the cold ischemia period has also been reported to be a confounding factor for biomarker expression and cooling to 2–8°C favoring the preservation of targets [28].

Concerning the choice of fixative to adequately preserve and stabilize tissue components for subsequent conventional histochemical and immunohistochemical analysis, 10% formalin (3.7% formaldehyde in water), though not perfect, is the preferred option. First of all, formalin has been used for more than a century and, in reality, all histological criteria and morphological features used by pathologists are based on this fixative. In addition, an incredibly large amount of archived paraffin blocks with formalin-fixed tissue have been generated over time to be used as valuable tools in clinical research, development of new methods, and quality assurance on a global basis supporting the choice of 10% formalin as the preferred fixative.

Formalin is a cross-linking fixative and, in short, the fixation is segmented in a three-step process with an initial penetration phase, followed by covalent binding phase, and ending with a cross-linking phase and final stabilization of peptides and proteins [32]. The penetration will proceed relatively fast, the covalent binding with moderate speed and cross-linking with slow speed [32]. The penetration phase depends on tissue type and size, but on average formalin is estimated to penetrate 1 mm per hour at room temperature. However, according to Baker, formalin penetrates with highest speed in the initial phase at a level of about 3.6 mm after one hour, but then the penetration rate slows down gradually and to double the depth of penetration to 7.2 mm, this takes 4 hours, to reach 14.4 mm depth, then 16 hours, etc. Following the penetration phase and formalin presence in the environment of proteins and peptides, the binding phase and cross-linking process is initiated. The cross-linking and stabilization phase is first estimated to be completed after 24–48 hours [33] or up to 7 days.[1] The critical concerns for formalin fixation focusing on IHC are both related to "under-fixation" and "over-fixation". The "under-fixation" addresses an inadequate preservation and stabilization of the target analyte with risk of loss or reduced expression during the subsequent tissue processing steps or archive storage, and in the end inducing a risk of false-negative results. This has been reported for many IHC biomarkers such as ER, ALK, and B-Raf proto-oncogene serine/threonine kinase (BRAF) [34–36].

Under-fixation and inadequate stabilization of proteins in general and of the target analyte can also cause a gradient in the IHC results being characterized by a positive result in the periphery of the tissue lesion and a gradually reduced or total loss of expression in the center of the tissue. This has frequently been observed and reported for IHC biomarkers such as mismatch repair proteins and can be caused by an adequate fixation and cross-linking in the periphery of the tissue, while the center is not stabilized and the exposure to alcohol and dehydration steps during tissue processing causes an extraction of the target proteins [37].

The "over-fixation" addresses the risk that the cross-linking of peptides induced by formalin fixation will change and mask the target antigens and the ability of the primary antibodies used for IHC to recognize these will be compromised. This risk has been significantly reduced by the discovery and use of epitope retrieval methods, and in diagnostic IHC over-fixation is not as critical and problematic as under-fixation [34, 38]. Especially heat-induced epitope retrieval (HIER) has been shown to be effective and an almost universal tool to restore antigens being blocked by the cross-linking process induced by formalin fixation. When HIER is applied by optimal conditions, antigens can be successfully demonstrated in tissue fixed for up to 7 weeks [39] and of central importance in allowing reproducible and reliable results for central diagnostic and predictive biomarkers such as ER and HER-2 in clinical samples fixed for 24–168 hours, which is the typical range in routinely processed samples for diagnostic pathology [40, 41]. However, not all antibodies will perform as expected after the application of HIER, and some antibodies will require epitope retrieval based on enzymatic digestion and for others no

---

[1] https://www.cambridge.org/core/services/aop-cambridge-core/content/view/28D59C7CCB0AD2136AD69948074907FC/S1551929500058491a.pdf/div-class-title-penetrationrates-of-formaldehyde-div.pdf (accessed on March 22, 2021).

epitope retrieval must be applied for an optimal performance. For antibodies requiring enzymatic digestion or no retrieval, the formalin fixation time will influence the demonstration of the target analyte and consequently the fixation time must be controlled and standardized for reproducible results [10, 39].

In order to evaluate the impact on the IHC result of central pre-analytical variables such as formalin fixation time, the pre-analytical TMA could then be designed to include tissue collected as fresh material with short (<0.5–1 hour) documented ischemic time and subsequently subjected to fixation in 10% neutral buffered formalin, e.g., 4–6 hours, 24–48 hours, and 168 hours reflecting the typical time range seen in a routine pathology laboratory. If the IHC test is also to be used on decalcified tissue, additional tissue processed with and without decalcification must be incorporated, as this can also affect the IHC result [10, 25, 27]. The IHC protocol and biomarker in focus can thereby be tested for the robustness of the core pre-analytical parameters and the impact can be critically evaluated.

In the era of the expanded use of biomarkers and IHC for precision medicine and in the central role of directing patient stratification and impacting clinical decisions, the focus on pre-analytical parameters and urge to strive for a process of standardization has become imminent. As biomarkers are multimodal and applied within both molecular and proteomic profiling, recommendations for "best practice" specimen handling must comprise most relevant and applied analysis performed on FFPE material and guidelines fit for IHC should not compromise the quality or possibility to perform molecular-based analysis for nucleic acids and vice versa. Several national and international standard institutes and organizations have provided detailed published guidelines and especially the jointly developed guidelines by ASCO/CAP for ER, PR, and HER-2 IHC testing have created a standard to comply with for pathology laboratories both for IHC breast cancer biomarkers and also as a general guideline. It has to be emphasized that compliance with the guidelines in many regions and countries is on voluntary basis unless the guidelines are part of an accreditation program such as ISO 15189 for Medical Laboratory Accreditation. The guidelines provided by ASCO/CAP for breast cancer biomarkers have also been adopted by the IASLC pathology committee for lung cancer and used to define guidelines for predictive IHC biomarkers such as PD-L1 [42]. In Table 2, see published recommendations for selected central pre-analytical steps for IHC and molecular analysis from different international and national authorities and organizations.

Table 2:   Guidelines for core pre-analytical procedures for tissues from international and national authorities.

Pre-analytical step	ASCO/CAP*	IASLC**	ISO/TC 212***
Biomolecule/method	ER-, PR-, HER2-IHC	PD-L1-IHC	Isolated DNA, RNA
Ischemic time	60 min. or less.	30 min. or less	Avoid or as short as possible
Type of fixative	10% NBF	10% NBF	10% NBF
Time in fixative	6–72 hours	6–48 hours	12–24 hours
Tissue thickness/ fixative ratio	5 mm/–	–/10:1	5 mm/4–10:1
Storage time/temp. for slides	6 weeks at RT[#]	8 weeks at RT[#]	Avoid/short at 2–8°C
Storage time/temp. for blocks	—	3 years/2–8°C or RT	–/2–8°C or RT

Notes: *American Society of Clinical Oncology (ASCO) and the College of American Pathologists (CAP).
**International Association for the Study of Lung Cancer (IASLC).
***European Committee for Standardization, ISO 20166.
[#]Room temperature.

Another relevant issue to address in the calibration phase relates to the storage conditions and age of cut slides and FFPE tissue blocks being both TMAs and whole tissue blocks. Several publications have confirmed a reduced level or even a total loss of antigenicity in stored unstained paraffin slides compared to freshly cut slides. The mechanism causing reduced antigenicity in stored paraffin slides is unidentified and multiple sources either occurring isolated or in combination such as high temperatures, moisture, light, under-fixation, and inadequate dehydration of tissue samples during the tissue processing steps may be the central factors causing a reduced immunoreactivity over time [43–45]. The publications and studies of antigen stability in pre-cut slides do show some diversity of the impact of storage, which can be contributed to the study design as well as materials and antibodies included. However, recent and controlled studies have shown that storage at room temperature for pre-cut slides will for up to 85% of routine IHC biomarkers induce a reduced IHC expression compared to freshly cut slides [46]. As seen in Table 2, the recommendation on storage of pre-cut sections at room temperature is 6–8 weeks to secure a stable antigen preservation. If longer storage time is needed, which especially is the situation for control slides for rare antigens, sections of TMAs to be used as internal tissue-tools for IHC method development and validation, or within a multicenter collaboration for research, storage at $-20°C$ or $-80°C$ for pre-cut slides should be the method of choice [47, 48].

Similar to storage conditions for pre-cut slides, the storage conditions can also have an impact on the IHC preservation and stability in archived tissue blocks. The vast majority of studies do indicate that the IHC biomarkers in general are preserved and proteins remain stable overtime in archived FFPE blocks. In many publications, long-term storage of paraffin blocks for 4–68 years did not reveal any impact of IHC staining as the expected IHC outcome in freshly cut sections from these stored blocks were obtained [27, 43, 44]. Of central importance, studies have indicated a stable expression in archived FFPE tissue for predictive IHC biomarkers such as data published by Ehinger *et al.* showing no loss for ER in up to 40-year-old tissue samples [49]. However, some studies as published by Combs *et al.* have shown that on average for the breast IHC biomarkers ER, HER-2, and Ki67, a loss of 10% of antigenicity was seen in FFPE blocks after 10 years' storage compared to the level demonstrated in fresh material [50]. Due to the diverging data, recommendations provided by national and international authorities frequently indicate a maximum time for storage of FFPE blocks for optimal IHC biomarker demonstration, especially with focus on predictive biomarkers as PD-L1 — see Table 2 [42].

2.4. *IHC analytical protocol optimization*

An IHC analytical protocol includes many variables and for each of these the laboratory must make a choice how to perform these, and the optimization process where different components and conditions are being tested is a key element for the laboratory to reach the goal to reproducibly generate high-quality IHC results.

An IHC analytical protocol consists of four core technical elements, which are antigen retrieval, primary antibody selection, detection system, and IHC instrumentation, see Figure 2.

The IHC protocol design can be segmented into two main categories. One category, named in-house or laboratory developed tests (LDT), are based on a primary antibody typically acquired as a concentrate, where the laboratory using serial sections of TMAs or individual tissue blocks

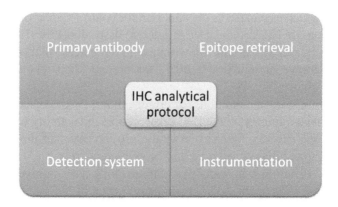

Figure 2:   The four core technical elements of an IHC protocol.

develops a protocol by testing several conditions for the core elements such as antigen retrieval methodology (no retrieval, enzymatic digestion, or HIER), HIER conditions (time, pH, and chemical composition of HIER buffer), primary antibody conditions (titer, incubation time, and diluent), detection system (type, incubation time, chromogen), and instrumentation type (if more are available in the laboratory). The other category is based on a ready-to-use (RTU) format of the primary antibody being applied in concordance with guidelines and protocol settings as described by the manufacturer concerning epitope retrieval, incubation time of RTU format, detection system, and instrumentation.

The primary antibodies to be used in either LDTs or as RTU formats are either monoclonal or polyclonal.

Monoclonal antibodies have traditionally become widely used because of their high specificity, consistency, and commercial availability for many target analytes. Monoclonal antibodies, produced in mice according to the classical *in vitro* hybridoma method or by recombinant DNA techniques, contain a single unique antibody molecule with unique specificity and affinity characteristics supporting the use and accuracy of IHC biomarkers [10, 51]. High specificity may occasionally be accompanied by a low affinity, consequently reducing the sensitivity of the antibody and the antibody has to be used in relatively concentrated form for optimal performance. Polyclonal antibodies, are typically being produced in rabbits by sequential booster immunization techniques for IHC to enhance the reactivity against the target antigen, and traditionally provide an increased sensitivity (avidity) compared to mouse monoclonal antibodies, as different antibody molecules in the antiserum react with more antigen sites at the target. As more antigen sites at the target are recognized by polyclonal antibodies, the risk of cross-reaction to other antigens is increased and specificity is reduced compared to monoclonal antibodies. However, polyclonal antibodies reacting with more antigen sites may minimize the deleterious impact of pre-analytical parameters such as formalin fixation, decalcification, and processing, thereby providing a more robust assay [10].

A third source of antibodies for IHC is rabbit monoclonal antibodies, which have gained popularity within diagnostic IHC since their introduction to the market [52]. The rabbit monoclonal antibodies possess the high specificity for murine monoclonal antibodies, being generated by the hybridoma technique and simultaneously also giving an increased avidity potentially being related to improved recognition of human antigens by the immune systems of rabbits compared to the murine counterpart.

The choice whether to select a mouse monoclonal antibody, a rabbit monoclonal antibody, or a polyclonal antibody must be determined by the individual laboratory, as the final performance and quality is highly dependent on the various pre-analytical and analytical parameters in the total test.

In Table 3, selected examples of widely used antibodies of different sources with focus on proportion of laboratories using these and performance as assessed by the external quality assessment program NordiQC are presented (www.nordiqc.org). The data are pooled regarding format as concentrate or RTU, with no attention to protocol settings, and thus only gives an overview of the market uptake and routine use of selected different antibody sources for diagnostics and associated quality outcome.

Epitope retrieval is another core element in the IHC analytical protocol and, as mentioned in Section 2.3, an indispensable tool to restore antigen sites that are blocked or structurally modified by the formalin fixation process. In the early period of diagnostic IHC in the 1970–1980s, epitope retrieval was purely based on application of proteolytic enzymes, but the discovery of HIER in 1991 by Shi revolutionized IHC as HIER was found effective to reproducibly restore a plethora of diagnostic relevant antigens in formalin-fixed lesions with minimal impact on fixation time [53–55]. HIER has virtually replaced proteolysis as the epitope retrieval method and a cautious estimate is applied for >95% of all IHC biomarkers in diagnostic pathology.

Table 3:   Sources, pass rates, and proportion of users for selected primary antibodies in NordiQC.

Antibody source and target	Performance and users	Primary intended use
	Antibody pass rate/Proportion users	Indication
Mouse monoclonal		
• CD20, clone L26	• 95%/96% (*n* = 161)	• Lymphoma subclassification
• Ki67, clone MIB1	• 90%/54% (*n* = 220)	• Proliferation index
• Cytokeratin 20, clone Ks20.8	• 91%/69% (*n* = 197)	• Origin of carcinoma
Rabbit monoclonal		
• ER, clone SP1	• 96%/65% (*n* = 235)	• Predictive marker, breast carc.
• Cyclin D1, clone SP4	• 94%/67% (*n* = 257)	• Lymphoma subclassification
• CDX2, clone EPR2764Y	• 89%/59% (*n* = 157)	• Origin of carcinoma
Rabbit polyclonal		
• S100, Z0311 & IR/IS/GA504*	• 88%/54% (*n* = 161)	• Origin of carcinoma
• PSA, A0562 & IR/IS/GA514*	• 89%/61% (*n* = 174)	• Origin of carcinoma

Note: *Product codes for the rabbit polyclonal antibody (Dako/Agilent).

In short, the efficiency of HIER is influenced by the heating time, heating temperature, pH, and chemical composition of the HIER buffer. The mechanism behind HIER is not fully clarified, but the combination of heating to >90°C, modifying the secondary and tertiary structures of proteins, and simultaneous hydrolysis of the formalin-fixed cross-linking methylene bridges most likely contributes to the revival of the immunoreaction. Other mechanisms are also involved such as removal of calcium ions, which strengthen the creation of methylene bridges and cross-linking by formalin fixation, but also changes in hydrophobic microenvironment and steric hindrance can be a co-factor as HIER can improve the performance of IHC antibodies in non-formalin-fixed and unfixed materials such as in cytology [56, 57].

There is generally an inverse correlation between heating temperature and heating time, as defined by the formula: HIER efficiency = Time × Temperature. If the heating is performed by pressure cooker at, e.g., 120°C, efficient epitope retrieval is completed after 2–3 minutes, compared to 10–20 minutes if the temperature is lowered to 90–100°C and processed in water bath or similar [54].

HIER can, as mentioned, be performed by different devices such as pressure cooker, water bath, micro oven, etc., but data from NordiQC clearly indicate that HIER for diagnostic IHC most commonly is performed as an integral step by fully automated IHC platforms such as BenchMark (Ventana/Roche), Omnis (Dako/Agilent), and BOND (Leica Biosystems). The efficiency of HIER in these platforms can be regulated by adjusting time and temperature.

The chemical composition and pH of the HIER buffer will also impact the efficiency of the epitope retrieval and no universal HIER buffer exists to provide optimal performance for all antibodies, therefore, the right choice regarding HIER settings in principle must be tailored to the individual antibody in question. Concerning pH of the HIER buffer, some antibodies will require low pH (1.0–2.0), others neutral pH (6.0–7.5) or high pH (8.0–10.0), but in general HIER performed at high pH will facilitate epitope retrieval for the vast majority of antibodies and is applied as the preferred option [58–60]. As mentioned, calcium ions interact with formalin during fixation and stabilize the cross-linking process, so removal of calcium ions is essential for effective antigen recovery. The use of ethylenediaminetetraacetic acid (EDTA) to chelate and efficiently remove calcium ions in the vicinity of methylene bridges and at the same time provide a basic pH in the range of 8.0–9.0 is more or less ubiquitously included in laboratory home-made and commercially available HIER buffers.

Proteolytic pre-treatment can be the preferred epitope retrieval method for few antibodies and optimal performance must be identified by selection of subtype of enzyme (e.g., Pepsin, Proteinase K, etc.), enzyme concentration, digestion time, and temperature. The reproducibility of IHC biomarkers requiring proteolysis is challenging and difficult to standardize, as optimal result will rely on tissue type and formalin fixation time, and the retrieval settings consequently must be adjusted to these variables, complicating the application in a routine pathology laboratory environment.

From data generated in NordiQC, a combined epitope retrieval based on HIER followed by proteolysis, or the other way around, might be superior to a "single" epitope retrieval method and a combined retrieval can improve both the analytical sensitivity and selectivity affecting the reaction pattern of the antibody, which consequently facilitates the readout of the IHC result. This was observed in the NordiQC assessment for WT-1, run 55, 2019, where the optimal result for clone 6F-H2 was obtained by a combined retrieval based on HIER followed by mild proteolysis providing a selective nuclear staining for WT-1, whereas HIER as single retrieval method induced an extensive co-existing cytoplasmic staining reaction hampering the readout.[2]

The outcome of an IHC test to a very high degree also depends on the choice of detection system, which is another core IHC analytical element. Per definition, a detection system is a tool or method to visualize the primary antibody attached to the target antigen. In diagnostic pathology, the primary antibody is unlabeled (except for fluorescence IHC in renal and skin auto-immune diseases) and the visualization of the primary antibody induced via application of labelled secondary antibodies reacting with the primary antibody creating a so-called indirect IHC method. The central aspects to take into account regarding detection systems is the level of sensitivity and specificity achievable, and the technical advances in IHC over the last decades have generated a broad range of commercially available detection systems to fulfill these two central aspects. Indirect polymer (dextran or micropolymers)-based detection systems are at present universally applied within diagnostic pathology laboratories and have virtually replaced biotin-based detection systems. Biotin-based detection

[2] https://www.nordiqc.org/downloads/assessments/117_77.pdf (assessed on March 15, 2021).

systems are still available but are used rarely, as the awareness of the risk of false-positive results related to the presence of endogenous biotin in many cell types has paved the way for the use of polymer systems at the expense of the biotin-based detection systems [61, 62].

Polymer detection systems are based on a compact polymer on which both secondary antibodies and enzyme molecules such as peroxidase or phosphatase are attached. The detection system can be applied within a two-step protocol based on incubation with the primary antibody followed by incubation with the polymer complex. Concerning level of sensitivity, this is also related to the level of complexity or more precisely number of steps included in the detection systems, and by inclusion of more steps with "linker" molecules the sensitivity can be increased. The choice of detection system will primarily depend on the affinity/avidity of the primary antibody and expression level of the target analyte and thereby expose if a highly sensitive method is required or a less complex method can be applied. It can be tempting to strive to use detection systems with maximal sensitivity to secure and optimize demonstration of the antigens, but a high level of complexity can reduce the reproducibility and signal-to-noise ratio hampering the final evaluation of the IHC result.

Figure 3 shows the level of complexity for different polymer-based detection systems and correlates this to the level of technical sensitivity of the system.

Number of steps					
5					Polymer + linker + TSA****
4				Polymer + 2 x linker***	
3			Polymer + linker**		
2		Polymer*			
		1	2	3	4
			Level of technical sensitivity		

Figure 3: Complexity of polymer-based detection system and level of technical sensitivity.

Notes: *Polymer or multimer (former generic name used for Dako/Agilent and Leica Biosystems products, latter for Ventana/Roche products).

**Rabbit anti-mouse or mouse anti-rabbit depending on primary antibody species.

***Rabbit anti-mouse and mouse anti-rabbit — sequence depending on primary antibody species.

****Tyramide signal amplification system.

The graded levels of technical sensitivity are not linear, but indicative as the final differences will depend on multiple factors such as polymer composition, sensitivity of linker molecules, application conditions of tyramide signal amplification, and sensitivity of chromogen [22, 63, 64].

In this context, inclusion of linker molecules can especially facilitate accessibility to certain epitopes, e.g., nuclear localized targets for polymer systems based on large molecules with tissue penetration challenges due to steric hindrances, linker molecules based on rabbit antibodies being more effective and sensitive compared to linker molecules based on mouse antibodies, and tyramide signal amplification can increase the level of sensitivity in the range 5–100 times depending on incubation times, reporter molecules, etc. [62]. These modalities give a wide span for the level of technical sensitivity possible and the impact will vary from system to system.

Looking at assessment data published from NordiQC, polymer-based detection systems with linker molecules (3- or 4-step methods) seem to be successful for most IHC biomarkers and as such a good choice as general backbone of IHC protocols for diagnostic use. This has been shown both for diagnostic biomarkers such as CD30, p40, PAX-8, SOX10, TTF-1, and also, which is of vital importance, for predictive biomarkers such as ALK with a very low expression level. Use of other detection systems can be relevant and mandatory for commercially available and validated RTU or CDx IHC assays such as HercepTest™ (Dako) and PATHWAY® (Ventana) for HER-2, both being developed and based on two-step polymer/multimer detection systems, or such as the CDx IHC SP142 assay (Ventana) for PD-L1 based on a multimer-based system with tyramide signal amplification. If a detection system is being changed for a validated IHC biomarker assay, the assay must be re-validated to confirm the accuracy.

2.5. *Instrumentation*

The combination of an almost consistently increasing request for IHC for diagnostics and a demand for standardization for patient safety has created a huge need for automation in IHC. Immunohistochemical assays are, as indicated, based on a multi-step procedure and in order to secure the reproducibility and reduce the manual handling of all steps in the analytical part such as performing epitope retrieval, application of reagents to the slides, washing and not least performing this identical for each slide processed, the access and use of automation in IHC is pivotal.

The automation of the analytical part of IHC is either based on a semi-automated or a full-automated system. For semi-automated systems, deparaffinization and HIER are performed in a dedicated and separate device such as a water bath, pressure cooker, or a pre-treatment module and when completed the slides are placed in another instrument to proceed and finalize the immunohistochemical analysis from primary antibody incubation till counterstaining, whereas full-automated systems include all analytical processes from deparaffinization till counterstaining. A survey conducted among the participants ($n = 349$) in the NordiQC EQA program for breast cancer immunohistochemistry indicated that 84% of the laboratories used a full-automated system compared to 14% using semi-automated systems, and only 2% performed the immunohistochemical assays manually [65]. For both semi- and full-automated systems, there are many different types offering different options and capabilities, but at the same time they also have different limitations and laboratories have to make a choice how to navigate and decide which system or systems will fulfill their needs. In this context it has to be mentioned that both based on personal experiences and data form the NordiQC EQA program generated in the years from 2003–2021, there is no universal best-in-class automated system for IHC — all have their pros and cons, and if possible, access to more than one automated system in the laboratory will be beneficial to secure a high and appropriate quality for all relevant immunohistochemical biomarkers.

The aspects to consider when selecting a solution for automation must address the following:

(1) Functionality — e.g., applicable for sample types such as FFPE, cytology, and frozen sections, batch or continuous loading options, epitope retrieval options (single retrieval only or combined as HIER + proteolysis), etc., (2) Workflow and workload — e.g., number of slides possible per run, day, and week, frequency and complexity of maintenance, etc., (3) Flexibility — e.g., modifications possible regarding protocol setup (sequence and conditions of retrieval settings, incubation times and temperature, options for single or multiplexing IHC, addition/removal of steps, etc.), access to a variety of reagents and antibodies both designed and developed for the system and from 3rd party, and finally (4) Costs — e.g., direct costs as price per unit, price per slide, costs for preventative

service, and indirect costs, e.g., related to accessories required, expected level of re-runs, down period, etc.

As mentioned above, the vast majority of laboratories performing diagnostic IHC are using full-automated instruments, with this also changing toward more or less closed systems. In open systems, the user, in its extreme setting, can modify all analytical protocol steps and select the reagents of choice such as HIER buffer, primary antibody (source, clone, titer, diluent), and detection system, whereas the integration and use of full-automated systems will reduce the options and flexibility. Typically, for full-automated systems central analytical protocol elements are restricted or fully locked and only reagents developed for the specific instrument can be applied, and as such the use of full-automation thereby will define the settings of three of the four central analytical protocol elements related to epitope retrieval, detection system, and instrumentation, leaving the primary antibody as main variable.

In this aspect, data from NordiQC clearly indicate that in addition to the use of full-automation, laboratories are switching toward an expanded usage of RTU primary antibodies developed and validated for the different full-automated systems at the expense of LDTs for diagnostic IHC. The shift to closed systems for IHC supports standardization and especially a central asset for inter- and intra-laboratory reproducibility of IHC, and has been a central factor to improve and maintain a high quality and accurate testing for predictive breast immunohistochemical biomarkers. As an example, it has been shown in the NordiQC EQA program for ER that in run B10, 2010 a total of 197 laboratories participated and an overall pass rate of 67% was seen. 52% of the laboratories used LTDs based on concentrated primary antibodies, while 48% used RTU systems obtaining a pass rate of 53% and 83%, respectively. In run B30, 2020 with 363 laboratories participating, the overall pass rate was improved to 92% and RTU systems were used by 86% with a pass rate of 94%, while the remaining 14% used LDTs giving a pass rate of 80%. Similar data have been observed and reported by NordiQC for the predictive biomarkers HER-2 and PD-L1, where the combination of full-automation and use of accurate RTU systems have played an important role in the standardization of IHC to support high quality precision testing [66].

On the downside for full-automated immunohistochemistry, publications and results from e.g. NordiQC have indicated that some primary antibodies will have different characteristics on different automated systems, impacting the final results [10, 67, 68]. This must be taken into account if a laboratory decides to transition their repertoire of immunohis-tochemical biomarkers from an existing automated system to another. The issue has in particular been observed when a mitigation is made from semi-automation using Autostainer (Dako/Agilent or Thermo Scientific™) to full-automation using Omnis (Dako/Agilent) or BenchMark (Ventana/Roche), where the chemical composition of reagents, sequence of reagent application, temperature, washing conditions, and/or mechanical handling of the sections within the full-automated system can have a detrimental effect on certain antibodies [10].

2.6. *Validation phase*

The validation phase is a pillar element in the lifecycle of quality manage-ment of IHC biomarkers. According to ISO 9000, validation is defined as "confirmation, through the provision of objective evidence, that require-ments for a specific intended use or application have been fulfilled" [69]. For pathology laboratories performing IHC for diagnostic use, the valida-tion of an IHC biomarker is synonymous to "technical" validation and based on data for the TPCs such as determination of analytical sensitivity and specificity (accuracy), demonstration of reportable range (limit of detection), analytical reproducibility, pre-analytical robustness, and where applicable data on readout accuracy and precision.

The term "specific intended use" listed by ISO 9000 can for IHC be translated to "purpose of IHC test" as outlined in Section 2.1 and "objec-tive evidence" defined by identification and use of appropriate TPCs to evaluate if the requirements have been fulfilled. The access and use of the tissue-tool box is in this phase an instrumental asset for the technical vali-dation process within the laboratory.

Initially in the validation process, the laboratory and user must define the purpose(s) of the IHC test, as this will influence the selection of tis-sues, lesions, or other relevant materials such as cell lines to be incorpo-rated in the test materials for the validation phase. At this stage, identification of tissues/materials with well-characterized and docu-mented pre-defined expression levels, equivalent to the tolerance LOD and expected reportable range, is essential. If the tissue/material does not represent the entire expected range of expression levels for the target ana-lyte, this might jeopardize the conclusions. This will especially be the outcome if the target analyte is expressed in a dynamic range from low to high, and if none or too few lesions with low level expression levels are included in the validation phase, the risk of false-negative results will occur when the validated test is implemented for diagnostic use [70].

There are no internationally accepted and standardized guidelines for the appropriate type and number of lesions to be tested during the IHC validation process to document the analytic accuracy and diagnostic potential. Suggestions from regulatory authorities as CAP, CLSI, and different expert committees indicate a range between 20 to 100 samples required with the number depending on parameters such as readout complexity of the IHC test (positive/negative vs. a semi-quantitative tests with different subgroups, e.g., HER-2 IHC), impact of IHC result being either diagnostic or predictive, and access to test material [23, 70, 71]. Irrespective of the number of samples, the samples must reflect the diagnostic purpose, the relevant expression levels of the target analyte, and comprise samples expected to be positive and negative. As shown in Figure 1, this can be accomplished by application of a TMA accuracy and TMA index comprising positive/negative samples and samples with the relevant and critical expression levels, e.g., being negative, weak and strong, respectively. For practical purposes, the TMA accuracy and TMA index can be merged and particularly possibly when a large number of samples are integrated in a TMA, thereby providing sufficient data to evaluate the performance and accuracy of the IHC assay with respect to both levels of intensity and extension, and expression being either positive or negative by defined readout criteria.

As mentioned above, the number of samples to be included in the validation phase to provide objective evidence that the IHC assay fulfills the needs required is not static and most critically is related to the risk of patient safety and impact on the treatment strategy in relation to the result obtained by the IHC assay. The risk evaluation of IHC assays has led to a segmentation and different nomenclature as proposed by Canadian Association of Pathologists (CAPACP), expert groups from ISIMM, and the International Quality Network for Pathology (IQN Path). Accordingly, to these societies a type-1 IHC assay is a diagnostic non-predictive IHC test used by the pathologist to classify a lesion and the IHC test typically applied within a panel with other IHC tests. As examples for type-1 IHC assays: p40, Cytokeratin 5, Napsin A, TTF1 as supportive tools to subclassify NSCLC as adenocarcinoma or squamous cell carcinoma.

The type-2 IHC assay is a "stand-alone" predictive IHC test where the pathologist performs the readout of the IHC result, but the result is used by the clinicians and oncologists to determine the treatment strategy for the patient. As examples for type-2 IHC assays: ER, HER-2, and PD-L1 in breast carcinoma and ALK, ROS1, and PD-L1 in NSCLC. The alignment of risk and IHC assay classification has also been addressed by The Food and Drug Administration (FDA) in the United States, categorizing IHC assays as Class I, II, and III. Class 1 is synonymous to type-1 IHC and Class 3 to type-2 IHC.

With respect to the risk evaluation and subclassification, type-1/ Class 1 IHC assays should be validated on minimum 20–40 clinical lesions, while type-2/Class 3 IHC assays should be validated on 40–100 lesions to comply with the guidelines from ASCO/CAP, CLSI, and other expert committees [26, 70–72]. The number of samples included in the validation will affect the probability that the test results observed in the validation phase reflect the results observed in "real diagnostic" settings, and a minimum of 10 positive and 10 negative lesions should be included in the validation set for virtually all non-predictive IHC biomarkers as indicated by CAP [71].

The accuracy of the IHC assay can be determined by several methods. The three most commonly used being:

(1) Comparing the new IHC assay results with a prior internally validated assay using same tissue set,
(2) Comparing the new IHC assay results with validated results from another laboratory using same tissue set,
(3) Comparing the new IHC assay results with an alternative validated non-IHC test (ISH, PCR, NGS, etc.), and
(4) Comparing the new IHC assay results with morphology, clinical data, and expected results.

In addition, and especially applicable when the laboratory is challenged to acquire appropriate number and types of relevant tissue samples for validation, to participate in external quality assessment (EQA) programs that offer proficiency testing using TMA samples with a sufficient number and proper selection of tissue samples, to provide evidence of test accuracy. The critical pre-analytical conditions for the samples used by the EQA provider must be similar to the conditions applied by the laboratory in order to exploit the data. Irrespective of the method applied to evaluate accuracy, a minimum of ≥90% concordance of the results between the new IHC assay and the one compared with as listed above must be obtained. For certain IHC biomarkers such as HER-2 IHC for breast cancer, a concordance of ≥95% can be required as directed by ASCO/CAP [26, 72]. As mentioned, the inclusion of more lesions supports the strength of evidence for the validated IHC assay being accurate for the intended purpose, and the use of 20 vs. 10, 40 vs. 20 lesions, etc., thus recommended by e.g., CAP. If the validation set comprises 10 lesions (5 positive and 5 negative), the overall concordance agreement (the level of agreement between two tests compared) reaches 90%, with 1 discordant result. The confidence interval (CI) of 95% (the range of values that has a 95% chance of including the "true" concordance) for the concordance agreement will be in the range of 57% to 100%. If the validation set is increased to 20 lesions, the overall concordance of 90% is obtained with two or fewer discordant results and a 95% CI of 69% to 98%, and finally using 40 lesions, 90% concordance is seen with four or fewer discordant results and a CI of 76–97% [26]. At this stage, the number of lesions and the need to create objective evidence is a balance on what is possible concerning access to lesions and must be decided by the laboratory from biomarker to biomarker.

For IHC biomarkers used for low-frequency target analytes, other approaches must be established, and this is also acknowledged by CAP in its accreditation checklist for IHC validation *the laboratory director determines that fewer validation cases are sufficient for a specific marker (e.g., a rare antigen or tissue), the rationale for that decision needs to be recorded* [73]. This is for instance highly relevant for IHC biomarkers identifying antigens/proteins induced by gene rearrangements such as ALK and ROS1 in NSCLCs, being observed in only 1–3% of NSCLCs, with similar challenges also seen for IHC for virus and microorganisms.

As supplement or alternative to identify and use in-house tissue and lesions, several vendors offer a wide range of commercially available TMAs of normal and malignant lesions and cell lines with varying expression levels of different target analytes, which can be used in the validation phase. Without any prioritization or being affiliated to the suggested vendors, the following selected sources for TMAs and cell lines are available and frequently cited in publications: US Biomax (www.biomax.us), Origene (www.origene.com), ProteoGenex (www.protegenex.com), and HistoCyte (www.histocyte.com).

As discussed in previous sections, RTU IHC systems are now being widely used for diagnostic IHC, and if these systems have been validated by the vendor and used according to package insert and within intended use, the validation process is significantly reduced and a confirmatory test design or so-called verification can be applied. The verification process is conducted in concordance with the guidelines for the RTU IHC system and will typically comprise IHC testing on 3–10 samples with characteristics as defined in the package insert [26].

A verification process of validated IHC tests is needed, when non-critical parameters of the IHC protocol are changed. Non-critical parameters being, e.g., antibody lot, antibody dilution change, same antibody clone but another vendor, etc. For these changes, the impact of the existing and modified settings should be compared by performing IHC on a minimum two positive and two negative lesions/tissues, and if accessible, it is preferable to include materials identified as ICAPCs.

Any change in the so-called critical parameters or in the core IHC technical elements will require a revalidation of the IHC assay. As examples: change in fixative, decalcification method, antibody clone, antigen retrieval settings, detection system, and/or IHC instrument will require a revalidation, as this can have a significant impact on the analytical sensitivity and specificity [26, 71]. Finally, if the use and purpose of a validated IHC test are changed, a revalidation typically is also warranted to ensure the test will accomplish the expected results for the new purpose. As example: the data extracted from NordiQC focusing on participants' IHC test accuracy for PD-L1 expression, to guide immune-oncology decision, revealed that laboratories mitigating successful IHC tests for PD-L1 in NSCLCs to be used also for TNBCs with the two different readout criteria associated for the two cancer types failed in up to 100% of the results generated.[3]

2.7. *Reproducibility phase*

Following the calibration and validation phase and having identified the best practice protocol for the new IHC biomarker, tools to secure and monitor the reproducibility of the IHC test must be identified. At this stage, a method transfer from development to implementation for clinical use takes place and it is of utmost importance for patient safety that the

---

[3] https://www.nordiqc.org/downloads/assessments/139_111.pdf (assessed on March 26, 2021).

expected level of analytical sensitivity/specificity defined through the development stage is maintained in each IHC test performed in the laboratory. The primary area to address is related to documenting that the IHC test was capable of demonstrating the target analyte in the clinically relevant expression range and LOD was accomplished, and in addition to confirm basic specificity indicating the right IHC test was applied.

Traditionally, positive and negative controls have been used in IHC as tools to monitor the test quality, but unfortunately and until recently the area has been neglected and has been the "missing link" in the approach to standardize IHC. This was underlined by the publication by Colascco performing a meta-analysis of 100 publications based on IHC concluding that in only 11% of these inclusions and descriptions of positive and negative controls were documented, and with right questioned the validity and reproducibility of data published [74].

The publications orchestrated by the international ad hoc expert panel for standardization of positive and negative controls in diagnostic IHC, followed by the evolutions series of "Quality Assurance for Clinical Immunohistochemistry in the Era of Precision Medicine" from 2014–2017 generated awareness of the importance of controls and provided details of terminology and specific recommendations with respect to IHC controls and what they are "controlling" [20, 21, 22, 24, 75].

It is highly recommended that these publications are used as inspiration and backbone for the design of IHC controls within the laboratory to evaluate the technical IHC test reproducibility and are a central asset in the quality management of diagnostic IHC.

In brief, the recommendations are based on the usage of a small TMA with 3–4 different normal human tissues being processed by the standard operating procedures as for clinical samples and the samples used in the development phase. The tissues included in the TMA must have been well characterized with respect to expected expression levels, cellular localization, and overall staining pattern for the antibodies to be evaluated by the TMA. If access to tissues with identified and validated iCAPC features, these should be preferred and included as being instrumental to evaluate that the expected performance and requirements for both LOD and technical specificity are fulfilled. Especially the inclusion of a positive control with LOD is critical, as e.g., data generated in NordiQC and also other proficiency schemes have revealed that in up to 90% of the insufficient results assessed, these have been characterized by too weak or false-negative IHC test results and caused by either poorly calibrated or less successfully performed methods [67, 68]. If the positive controls "only" have high expression levels, no information on the level of analytical sensitivity and relevant LOD can be extracted with risk of false-negative results. The importance of positive controls with LOD to guide the level of analytical sensitivity have been addressed extensively by NordiQC and for the vast majority of the IHC biomarkers evaluated information of recommendations of positive controls with LOD can be found for the individual biomarkers at www.nordiqc.org. Recently, ASCO/CAP in the 2020 updated guidelines for ER and PR specifically included recommendations for positive tissue controls as tonsil and uterine cervix for the evaluation of LOD are pivotal for correct treatment decision [70].

The approach to identifying iCAPCs and integrating these in daily quality control for IHC tests is important for the standardization of IHC, but for certain IHC biomarkers no tissue with iCAPC capabilities can be identified or the iCAPC identified would not be reliable to monitor LOD and IHC test reproducibility. This is in particular critical for CDx assays with a demand for a threshold-based readout to separate a negative and a positive IHC test result, and these challenges have been observed especially for PD-L1 and HER-2 IHC testing as well as for other "quantitative" IHC tests such as Ki67, and for these no or only limited access to control material with reliable and validated LOD capabilities and iCAPC potential exist.

Many ongoing studies and projects focus on the identification and validation of non-tissue-based controls such as cell lines, histoids, xenografts, peptides, etc. to be used as reliable controls typically offering a "dynamic range" for the expression for the target analyte. For both tissues and non-tissue-based iCAPCs, the verification of the level of analytical sensitivity and LOD is based on descriptive characteristics typically regarding the proportion of cells being positive and the intensity of the staining reaction. These controls serve as "descriptive" calibrators as the concentration of the target analyte is unknown, although at a low level, and cannot be measured according to a national or international standard with defined units. Initiatives to create reference standards with accurate measurement traceability for IHC are emerging and might be valuable and superior to existing controls to evaluate IHC reproducibility and LOD levels [76, 77]. These systems are based on microbeads coated with single peptides of the target analyte, e.g., ER and a series of microbead pellets with a different range and concentration of the target peptides (e.g., 1,000–1,000,000 units) being applied on a glass slide. The LOD of the

IHC test can subsequently be determined by the microbead pellet with lowest concentration being positive, and in the end optimally verified by image analysis giving an exact LOD. Recent publications have evaluated the concept and compared the potential of microbeads as calibrations with conventional tissue controls to evaluate IHC test quality for ER and indicated a superior accuracy of the microbead system to identify IHC protocols, giving too low analytical sensitivity [78]. The combination of image analysis and control material with dynamic expression levels of a target analyte might also be a promising tool to evaluate reproducibility of IHC tests, and feasibility studies focusing on Ki67 conclude that image analysis and cell lines with characterized Ki67 index might be more accurate for the technical evaluation of the analytical component of a Ki67 test compared to the use of "human expert evaluation" and conventional tissue controls [79].

Finally, to evaluate the reproducibility of IHC tests, it has to be emphasized that it is not only a question of selecting the right control material, it is also essential how the control materials are used, and in this aspect the use of on-slide controls are highly recommended by both IHC expert groups and also ASCO/CAP to have the highest assurance that the IHC result is precise from test to test [24, 80]. IHC controls can be applied as a separate slide and termed "run" or "batch" controls and stained separately to the diagnostic samples with same protocol, whereas on-slide controls are mounted at the same slide as the diagnostic sample. In the era when IHC is mainly performed within full-automated systems with separate and individual slide handling, only the use of on-slide controls can document that both the external control and the patient sample are processed by identical test conditions, while this by nature is not possible if batch controls are used.

References

[1] Bellizzi, A. M. (2020). An algorithmic immunohistochemical approach to define tumor type and assign site of origin. *Adv. Anat. Pathol.* 27(3): 114–163.
[2] Stelow, E. B. and Yaziji, H. (2018). Immunohistochemistry, carcinomas of unknown primary, and incidence rates. *Semin. Diagn. Pathol.* 35(2): 143–152.
[3] Gu, J. and Taylor, C. R. (2014). Practicing pathology in the era of big data and personalized medicine. *Appl. Immunohistochem. Mol. Morphol.* 22(1): 1–9.

[4] Taylor, C. R. (2014). Predictive biomarkers and companion diagnostics. The future of immunohistochemistry: "*In situ* proteomics", or just a "stain"? *Appl. Immunohistochem. Mol. Morphol.* 22(8): 555–561.

[5] Strimbu, K. and Tavel, J. A. (2010). What are biomarkers? *Curr. Opin. HIV AIDS* 5(6): 463–466.

[6] Jørgensen, J. T. (2016). Companion and complementary diagnostics: Clinical and regulatory perspectives. *Trends Cancer* 2(12): 706–712.

[7] Inamura, K. (2018). Update on immunohistochemistry for the diagnosis of lung cancer. *Cancers (Basel)* 10(3): 72.

[8] Cottrell, T. R. and Taube, J. M. (2018). PD-L1 and emerging biomarkers in immune checkpoint blockade therapy. *Cancer J.* 24(1): 41–46.

[9] Taylor, C. R. (2000). The total test approach to standardization of immuno-histochemistry. *Arch. Pathol. Lab. Med.* 124(7): 945–951.

[10] Education Guide. *Immunohistochemical Staining Methods*. Sixth Edition, Dako. Taylor CR. Chapter 4:46–59 Nielsen S. Available at https://www.agilent.com/cs/library/technicaloverviews/public/08002_ihc_staining_methods.pdf (accessed on February 23, 2021).

[11] Lin, F. and Chen, Z. (2014). Standardization of diagnostic immunohisto-chemistry: Literature review and geisinger experience. *Arch. Pathol. Lab. Med.* 138(12): 1564–1577.

[12] Cates, J. M. and Troutman, K. A. Jr. (2015). Quality management of the immunohistochemistry laboratory: A practical guide. *Appl. Immunohistochem. Mol. Morphol.* 23(7): 471–480.

[13] Luk, P. P., Selinger, C. I., Mahar, A., and Cooper, W. A. (2018). Biomarkers for ALK and ROS1 in lung cancer: Immunohistochemistry and fluorescent *in situ* hybridization. *Arch. Pathol. Lab. Med.* 142(8): 922–928.

[14] Cutz, J. C., Craddock, K. J., Torlakovic, E. *et al.* (2014). Canadian anaplas-tic lymphoma kinase study: A model for multicenter standardization and optimization of ALK testing in lung cancer. *J. Thorac. Oncol.* 9(9): 1255–1263.

[15] Mino-Kenudson, M., Chirieac, L. R., Law, K. *et al.* (2010). A novel, highly sensitive antibody allows for the routine detection of ALK-rearranged lung adenocarcinomas by standard immunohistochemistry. *Clin. Cancer Res.* 16(5): 1561–1571.

[16] Torlakovic, E., Lim, H. J., Adam, J. *et al.* (2020). "Interchangeability" of PD-L1 immunohistochemistry assays: A meta-analysis of diagnostic accu-racy. *Mod. Pathol.* 33(1): 4–17.

[17] Udall, M., Rizzo, M., Kenny, J. *et al.* (2018). PD-L1 diagnostic tests: A systematic literature review of scoring algorithms and test-validation metrics. *Diagn. Pathol.* 13(1): 12.

[18] Hirsch, F. R., McElhinny, A., Stanforth, D. *et al.* (2017). PD-L1 immuno-histochemistry assays for lung cancer: Results from phase 1 of the

blueprint PD-L1 IHC assay comparison project. *J. Thorac. Oncol.* 12(2): 208–222.

[19] Tsao, M. S., Kerr, K. M., Kockx, M. *et al.* (2018). PD-L1 immunohisto-chemistry comparability study in real-life clinical samples: Results of blueprint phase 2 project. *J. Thorac. Oncol.* 13(9): 1302–1311.

[20] Cheung, C. C., D'Arrigo, C., Dietel, M. *et al.* (2017). Evolution of quality assurance for clinical immunohistochemistry in the era of precision medi-cine: Part 1: Fit-for-purpose approach to classification of clinical immuno-histochemistry biomarkers. *Appl. Immunohistochem. Mol. Morphol.* 25(1): 4–11.

[21] Cheung, C. C., D'Arrigo, C., Dietel, M. *et al.* (2017). Evolution of quality assurance for clinical immunohistochemistry in the era of precision medi-cine: Part 4: Tissue tools for quality assurance in immunohistochemistry. *Appl. Immunohistochem. Mol. Morphol.* 25(4): 227–230.

[22] Torlakovic, E. E., Cheung, C. C., D'Arrigo, C. *et al.* (2017). Evolution of quality assurance for clinical immunohistochemistry in the era of precision medicine — part 2: Immunohistochemistry test performance characteris-tics. *Appl. Immunohistochem. Mol. Morphol.* 25(2): 79–85.

[23] Torlakovic, E. E., Cheung, C. C., D'Arrigo, C. *et al.* (2017). Evolution of quality assurance for clinical immunohistochemistry in the era of precision medicine. Part 3: Technical validation of immunohistochemistry (IHC) assays in clinical IHC laboratories. *Appl. Immunohistochem. Mol. Morphol.* 25(3): 151–159.

[24] Torlakovic, E. E, Nielsen, S., Francis, G. *et al.* (2015). Standardization of positive controls in diagnostic immunohistochemistry: Recommendations from the International Ad Hoc Expert Committee. *Appl. Immunohistochem. Mol. Morphol.* 23(1): 1–18.

[25] Engel, K. B. and Moore, H.M. (2011). Effects of pre-analytical variables on the detection of proteins by immunohistochemistry in formalin-fixed, paraffin-embedded tissue. *Arch. Pathol. Lab. Med.* 135(5): 537–543.

[26] Ibarra, J. A., Rogers, L. W., Kyshtoobayeva, A., and Bloom, K. (May 2010). Fixation time does not affect the expression of estrogen receptor. *Am. J. Clin. Pathol.* 133(5): 747–755.

[27] Bass, B. P., Engel, K. B., Greytak, S. R., and Moore, H. M. (2014). A review of pre-analytical factors affecting molecular, protein, and morpho-logical analysis of formalin-fixed, paraffin-embedded (FFPE) tissue: How well do you know your FFPE specimen? *Arch. Pathol. Lab. Med.* 138(11): 1520–1530, doi:10.5858/arpa.2013-0691-RA.

[28] Yildiz-Aktas, I. Z., Dabbs, D. J., and Bhargava, R. (2012). The effect of cold ischemic time on the immunohistochemical evaluation of estrogen receptor, progesterone receptor, and HER2 expression in invasive breast carcinoma. *Mod. Pathol.* 25: 1098–1105.

[29] Lee, A. H., Key, H. P., Bell, J. A. *et al.* (2014). The effect of delay in fixation on HER2 expression in invasive carcinoma of the breast assessed with immunohistochemistry and *in situ* hybridisation. *J. Clin. Pathol.* 67: 573–575.

[30] Khoury, T. (2018). Delay to formalin fixation (Cold ischemia time) effect on breast cancer molecules. *Am. J. Clin. Pathol.* 149(4): 275–292.

[31] Neumeister, V. M., Anagnostou, V., Siddiqui, S. *et al.* (2012). Quantitative assessment of effect of pre-analytic cold ischemic time on protein expression in breast cancer tissues. *J. Natl. Cancer Inst.* 104: 1815–1824.

[32] Dapson, R. W. (2010). Mechanisms of action and proper use of common fixatives. In *Antigen Retrieval Immunohistochemistry Based Research and Diagnostics*, Shi, S.-R. and Taylor, C. R. (eds.), Hoboken, NJ: John Wiley, pp. 195–217.

[33] Werner, M., Chott, A., Fabiano, A., and Battifora, H. (2000). Effect of formalin tissue fixation and processing on immunohistochemistry. *Am. J. Surg. Pathol.* 24(7): 1016–1019.

[34] Goldstein, N. S., Ferkowicz, M., Odish, E. *et al.* (2003). Minimum formalin fixation time for consistent estrogen receptor immunohistochemical staining of invasive breast carcinoma. *Am. J. Clin. Pathol.* 120: 86–92.

[35] Miller, R., Thorne-Nuzzo, T., Loftin, I., McElhinny, A., Towne, P., and Clements, J. (2020). Impact of pre-analytical conditions on the antigenicity of lung markers: ALK and MET. *Appl. Immunohistochem. Mol. Morphol.* 28(5): 331–338.

[36] Dvorak, K., Aggeler, B., Palting, J., McKelvie, P., Ruszkiewicz, A., and Waring, P. (2014). Immunohistochemistry with the anti-BRAF V600E (VE1) antibody: Impact of pre-analytical conditions and concordance with DNA sequencing in colorectal and papillary thyroid carcinoma. *Pathology* 46(6): 509–517.

[37] Remo, A., Fassan, M., and Lanza, G. (2016). Immunohistochemical evaluation of mismatch repair proteins in colorectal carcinoma: The AIFEG/GIPAD proposal. *Pathologica* 108: 104–109.

[38] Arber, D. A. (2002). Effect of prolonged formalin fixation on the immunohistochemical reactivity of breast markers. *Appl. Immunohistochem. Mol. Morphol.* 10(2): 183–186.

[39] Webster, J. D., Miller, M. A., Dusold, D., and Ramos-Vara, J. (August 2009). Effects of prolonged formalin fixation on diagnostic immunohistochemistry in domestic animals. *J. Histochem. Cytochem.* 57(8): 753–761.

[40] Tong, L. C., Nelson, N., Tsourigiannis, J., and Mulligan, A. M. (April 2011). The effect of prolonged fixation on the immunohistochemical evaluation of estrogen receptor, progesterone receptor, and HER2 expression in invasive breast cancer: A prospective study. *Am. J. Surg. Pathol.* 35(4): 545–552.

[41]   Kai, K., Yoda, Y., Kawaguchi, A., Minesaki, A., Iwasaki, H., Aishima, S., and Noshiro, H. (February 26, 2019). Formalin fixation on HER-2 and PD-L1 expression in gastric cancer: A pilot analysis using the same surgical specimens with different fixation times. *World J. Clin. Cases* 7(4): 419–430.

[42]   Lantuejoul, S., Sound-Tsao, M., Cooper, W. A. *et al.* (2020). PD-L1 testing for lung cancer in 2019: Perspective from the IASLC pathology committee. *J. Thorac. Oncol.* 15(4): 499–519.

[43]   Manne, U., Myers, R. B., Srivastava, S., and Grizzle, W. E. (1997). Re: Loss of tumor marker-immunostaining intensity on stored paraffin slides of breast cancer. *J. Natl. Cancer Inst.* 89(8): 585–586.

[44]   Vis, A. N., Kranse, R., Nigg, A. L., and van der Kwast, T. H. (2000). Quantitative analysis of the decay of immunoreactivity in stored prostate needle biopsy sections. *Am. J. Clin. Pathol.* 113(3): 369–373, doi:10.1309/ CQWY-E3F6-9KDN-YV36.

[45]   Wester, K., Wahlund, E., Sundström, C. *et al.* (2000). Paraffin section storage and immunohistochemistry. Effects of time, temperature, fixation, and retrieval protocol with emphasis on p53 protein and MIB1 antigen. *Appl. Immunohistochem. Mol. Morphol.* 8(1): 61–70.

[46]   Ramos-Vara, J. A., Webster, J. D., DuSold, D., and Miller, M. A. (2014). Immunohistochemical evaluation of the effects of paraffin section storage on biomarker stability. *Vet. Pathol.* 51(1): 102–109.

[47]   Andeen, N. K., Bowman, R., Baullinger, T., Brooks, J. M., and Tretiakova, M.S. (2017). Epitope preservation methods for tissue microarrays: Longitudinal prospective study. *Am. J. Clin. Pathol.* 148(5): 380–389.

[48]   Forse, C. L., Pinnaduwage, D., Bull, S. B., Mulligan, A. M., and Andrulis, I. L. (2019). Fresh cut versus stored cut paraffin-embedded tissue: Effect on immunohistochemical staining for common breast cancer markers. *Appl. Immunohistochem. Mol. Morphol.* 27(3): 231–237.

[49]   Ehinger, A., Bendahl, P.-O., Rydén, L., Fernö, M., and Alkner, S. (2018). Stability of oestrogen and progesterone receptor antigenicity in formalin-fixed paraffin-embedded breast cancer tissue over time. *APMIS* 126: 746–754.

[50]   Combs, S. E., Han, G., Mani, N., Beruti, S., Nerenberg, M., and Rimm. D. L. (2016). Loss of antigenicity with tissue age in breast cancer. *Lab. Invest.* 96(3): 264–269.

[51]   Kohler, G., Howe, S. C., and Milstein, C. (1976). Fusion between immuno-globulin-secreting and non-secreting myeloma cell lines. *Eur. J. Immunol.* 6: 292–295.

[52]   Rossi, S. *et al.* (2005). Rabbit monoclonal antibodies. A comparative study between a novel category of immunoreagents and the corresponding mouse monoclonal antibodies. *Am. J. Clin. Pathol.* 124(2): 295–302.

[53]   Shi, S. R., Key, M. E., and Kalra, K. L. (1991). Antigen retrieval in forma-lin-fixed, paraffin-embedded tissues: An enhancement method for

immunohistochemical staining based on microwave oven heating of tissue sections. *J. Histochem. Cytochem.* 39(6): 741–748.

[54] Shi, S.-R., Cote, R. J., and Taylor, C. R. (1997). Antigen retrieval immuno-histochemistry: Past, present, and future. *J. Histochem. Cytochem.* 45(3): 327–343.

[55] Boenisch, T. (September 2005). Effect of heat-induced antigen retrieval following inconsistent formalin fixation. *Appl. Immunohistochem. Mol. Morphol.* 13(3): 283–286.

[56] Jain, D., Nambirajan, A., Borczuk, A. *et al.* (2019). Immunocytochemistry for predictive biomarker testing in lung cancer cytology. *Cancer Cytopathol.* 127(5): 325–339.

[57] Morgan, J. M., Navabi, H., and Jasani, B. (June 1997). Role of calcium chelation in high-temperature antigen retrieval at different pH values. *J. Pathol.* 182(2): 233–237.

[58] Kajiya, H., Takekoshi, S., Takei, M. *et al.* (2009). Selection of buffer pH by the isoelectric point of the antigen for the efficient heat-induced epitope retrieval: Re-appraisal for nuclear protein pathobiology. *Histochem. Cell Biol.* 132(6): 659–667.

[59] Krenacs, L., Krenacs, T., Stelkovics, E., and Raffeld, M. (2010). Heat-induced antigen retrieval for immunohistochemical reactions in routinely processed paraffin sections. *Methods Mol. Biol.* 588: 103–119.

[60] Shi, S. R., Imam, S. A., Young, L., Cote, R. J., and Taylor, C. R. (1995). Antigen retrieval immunohistochemistry under the influence of pH using monoclonal antibodies. *J. Histochem. Cytochem.* 43(2): 193–201.

[61] Gown, A. M. (September 1, 2016). Diagnostic immunohistochemistry: What can go wrong and how to prevent it. *Arch. Pathol. Lab. Med.* 140(9): 893–898.

[62] Shojaeian, S., Lay, N. M., and Zarnani, A.-H. (December 17, 2018). *Detection Systems in Immunohistochemistry, Immunohistochemistry — The Ageless Biotechnology.* Streckfus, C. F. (ed.), IntechOpen. Available at https://www.intechopen.com/books/immunohistochemistry-the-ageless-biotechnology/detection-systems-in-immunohistochemistry (accessed on March 15, 2021).

[63] Torlakovic, E. E., Nielsen, S., Vyberg, M. *et al.* (2015). Getting controls under control: The time is now for immunohistochemistry. *J. Clin. Pathol.* 68: 879–882.

[64] Seidl, M., Weinhold, B., Jacobsen, L., Rasmussen, O. F., Werner, M., and Aumann, K. (December 2020). Critical assessment of staining properties of a new visualization technology: A novel, rapid and powerful immunohisto-chemical detection approach. *Histochem. Cell Biol.* 154(6): 663–669.

[65] https://www.nordiqc.org/downloads/documents/127.pdf (accessed on March 21, 2021).

[66] Vyberg, M., Nielsen, S., Roge, R., Sheppard, B., Ranger-Moore, J., Walk, E. *et al.* (September 2013). HER2-based treatment decisions in breast cancer (BC): Test accuracy and its clinical, economic and social impact. *Eur. J. Cancer.* 49(Suppl. 2): S203–S203. 981.

[67] Vyberg, M. and Nielsen, S. (2016). Proficiency testing in immunohistochemistry — experiences from Nordic Immunohistochemical Quality Control (NordiQC). *Virchows Arch.* 468(1): 19–29.

[68] Nielsen, S. (2015). External quality assessment for immunohistochemistry: Experiences from NordiQC. *Biotech. Histochem.* 90(5): 331–340.

[69] International Organization for Starndardization. (2015). ISO 9000:2015(en), Quality management systems—fundamentals and vocabulary. Available at https://www.iso.org/obp/ui/#iso:std:iso:9000:en (accessed on February 24, 2021).

[70] Allison, K. H., Hammond, M. E. H, Dowsett, M. *et al.* (2020). Estrogen and progesterone receptor testing in breast cancer: ASCO/CAP guideline update. *J. Clin. Oncol.* 38(12): 1346–1366.

[71] Fitzgibbons, P. L., Bradley, L. A., Fatheree, L. A. *et al.* (2014). Principles of analytic validation of immunohistochemical assays: Guideline from the college of American pathologists pathology and laboratory quality center. *Arch. Pathol. Lab. Med.* 138: 1432–1443.

[72] Wolff, A. C., Hammond, M. E. H, Allison, K.H. *et al.* (2018). Human epidermal growth factor receptor 2 testing in breast cancer: American society of clinical oncology/college of American pathologists clinical practice guideline focused update. *Arch. Pathol. Lab. Med.* 142(11): 1364–1382.

[73] College of American Pathologists, 2017 Anatomic pathology checklist, ANP.22750 — Antibody Validation.

[74] Colasacco, C., Mount, S., and Leiman, G. (2011). Documentation of immunocytochemistry controls in the cytopathologic literature: A meta-analysis of 100 journal articles. *Diagn. Cytopathol.* 39(4): 245–250.

[75] Torlakovic, E. E., Francis, G., Garratt, J. *et al.* (2014). International ad hoc expert panel. Standardization of negative controls in diagnostic immunohistochemistry: Recommendations from the international ad hoc expert panel. *Appl. Immunohistochem. Mol. Morphol.* 22(4): 241–252.

[76] Sompuram, S., Vani, K., Schaedle, A. *et al.* (2018). Quantitative assessment of immunohistochemistry laboratory performance by measuring analytic response curves and limits of detection. *Arch. Pathol. Lab. Med.* 142: 851–862.

[77] Bogen, S. A. (2019). A root cause analysis into the high error rate in clinical immunohistochemistry. *Appl. Immunohistochem. Mol. Morphol.* 27(5): 329–338.

[78] Torlakovic, E. E., Sompuram, S., Vani, K., Wang, L. *et al.* (February 15, 2021). Introduction and clinical validation of metrology standards for

immunohistochemistry (IHC); New tool for standardization of estrogen receptor (ER) IHC assay in breast cancer. *Cancer Res.* 81(4 Suppl.): PS5–PS46.

[79] Røge, R., Nielsen, S., Riber-Hansen, R. *et al.* (February 2021). Image analyses assessed cell lines as potential performance controls of Ki-67 immunostained slides. *Appl. Immunohistochem. Mol. Morphol.* 29(2):95–98.

[80] Cheung, C. C., Taylor, C. R., and Torlakovic, E. E. (May/June 2017). An audit of failed immunohistochemical slides in a clinical laboratory: The role of on-slide controls. *Appl. Immunohistochem. Mol. Morphol.* 25(5): 308–312.

https://doi.org/10.1142/9781800611399_0005

Chapter 5

Measuring Ki-67 in Breast Cancer: Past, Present, and Future

Andrew Dodson

UK NEQAS for Immunocytochemistry & In-Situ Hybridisation, UK

adodson@ukneqasiccish.org

1. Introduction and Historical Context

The identification and characterization of the first monoclonal antibody reactive against the Ki-67 antigen was made by Johannes Gerdes in the early 1980s. At the time of his discovery, Gerdes was a Ph.D. student in the laboratories of Harald Stein, a German pathologist and researcher who was then studying Hodgkin's lymphoma. Gerdes was charged with the task of finding monoclonal antibodies that reacted selectively with the dividing cells of this tumor type. The antibody producing clone was growing in the 67th well of the multi-well tissue culture plate and Stein's laboratories were located in Kiel in Germany, hence the antibody producing clone was named Ki-67 [1]. Figure 1 shows one of the paper's illustrative photomicrographs with this example being of Ki-67's staining pattern in tonsil.

Gerdes quickly realized the antibody's specificity extended beyond proliferating nuclei of tumor cells in Hodgkin's disease and that it appeared to identify a nuclear protein present in all dividing cells as he and his fellow authors demonstrated in their first papers characterizing the antibody [2, 3]. In their second paper, the group demonstrated the protein

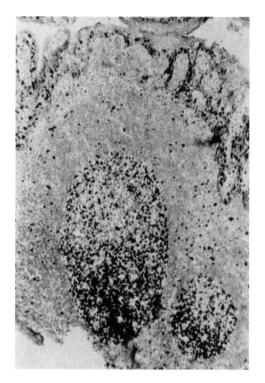

Figure 1:   Frozen-section of tonsil stained using the original Ki-67 antibody clone in an immunoperoxidase technique. In follicle centers the majority of centroblasts in the dark zone and a smaller proportion of centrocytes in the light zone are stained, in contrast few sparsely scatted Ki-67 positive cells are seen in the inter-follicular zone. Within the epithelial compartment basal cells show more prevelent reactivity compared to the superficial layers. Note how similar the staining appears to that we would expect to see in a modern-day ICC preparation. Illustration taken from the first paper describing the discovery of the antibody. Adapted with permission from Ref. [1]. Copyright John Wiley and Sons.

was present in nuclei of cells in S, G2, and M phase but absent in G0. They also showed that cells in G1 which had yet to undergo mitosis were Ki-67-negative and that they gained Ki-67 expression after mitosis had occurred [3].

The significance of this was not lost on the group, as demonstrated by this excerpt from the discussion section of their 1983 paper, *Ki-67 might become an important tool for reliably and quickly determining the proliferation rates of malignant tumors. The exact determination of the percentage of proliferating cells in neoplasms appears to be of prognostic value and may also become important for the choice of appropriate treatment* [2].

In the more than three-and-a-half decades that have passed since those statements were written, Ki-67 has indeed become one of our most widely applied tools for assessing cellular proliferation as witnessed by the fact that it is the subject of approximately 1,800 research articles published year-on-year [4]. Many of them dealing with its use as a prognostic tool.

During much of those 35+ years, the exact function of the Ki-67 protein in the cell cycle was unknown. Indeed, in 2002 in an article published nearly 20-years after its discovery, the authors were able to state that *Given the vast amount of information we have on the structure, localization and regulation of pKi-67, it is extraordinary that so little is known about the function of this protein* [5]. It was 14 years later in 2016 that the situation changed dramatically with the publication by a group based in Austria and Germany of a ground-breaking seminal paper conclusively identifying Ki-67 as a protein with a unique and central role in cell-division [6].

The group used a variety of molecular biology-based experiments to first identify *MKI67* as the gene responsible for producing a protein (Ki-67) that, when lost, led to clustering of chromosomes at mitosis as opposed to their normal dispersion and separation. Knock-out models produced using small interfering RNA constructs proved that Ki-67 was absolutely essential to the separation of daughter chromosomes. Next, they attempted to map the function of reducing chromosome surface adhesiveness to specific regions on the Ki-67 protein. Ki-67 is a very large, highly electrically charged, largely unfolded protein. Its N-terminus contains a phosphopeptide-binding Forkhead-associated (FHA) region, the central region is composed largely of tandem repeats, and the C-terminus is enriched for leucine and arginine (LR) pairs. The group were surprised to find that very little of the protein's specific structure was required to maintain its ability to function; one part which was needed was the LR domain, and if this was missing, the protein did not retain any affinity for the chromosomal surface whatsoever. However, loss of the N-terminal FHA region had almost no effect on function. Extensive further experiments with a variety of Ki-67-like constructs led to the conclusion that the Ki-67 proteins' amphiphilic long-chain structure, with its chromatin attractive LR C-terminal and cytosol-attractive N-terminal separated by a long straight region, was key to the way it functioned, and that this was

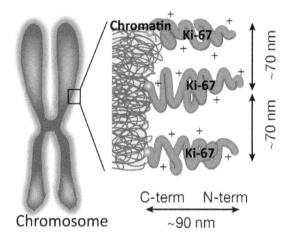

Figure 2: Ki-67 protein acts as a "non-stick" surface coating surrounding individual chromosomes at mitosis, and so allows them to separate along the mitotic spindle. The Ki-67 molecules are depicted here as loosely coiled protien "strings" with their C-terminal leucine-argine rich 'head's closely attached to the chromosomal chromatin. As the density of Ki-67 molecules increases their positive electrical charge causes them the open-out into straight "bristles" [2]. Adapted with permission from the Ref. [6]. Copyright Springer Nature.

very similar to surface-active agents (surfactants). Thus, Ki-67 protein molecules form an effective repulsive barrier by covering the whole surface of mitotic chromosomes at a density which causes the extension of the protein into structures analogous to the bristles of a brush (Figure 2).

2. Ki-67 as a Prognostic Marker

In a comprehensive review of the evidence around Ki-67 as a prognostic marker in breast cancer, Luporsi and her colleagues identified 17 randomized clinical trials (RCT's) with results published between 1990 and 2010 [7]. Multivariate analyses of those studies which had secondary central analysis of Ki-67 showed the marker to be an independent prognostic factor for disease-free survival (HR 1.05–1.72). The level of evidence was judged to be I-B using the definition of Simon [8], i.e., There are consistent results from one or more trials, but the evidence was from RCTs not specifically designed to assess the utility of the biomarker, with the samples being centrally analyzed after the study closed.

Since the publication of Luporsi's survey, results of a number of additional RCTs using Ki-67 as a prognostic marker have been published. Most significantly in respect of this review is the Peri-Operative Endocrine Therapy — Individualizing Care (POETIC) phase 3 RCT in early hormone sensitive breast cancer that published its results in 2020 [9]. The trial examined the hypothesis that Ki-67 could be used to predict outcome after short duration pre-operative endocrine therapy and could therefore be helpful in accurately selecting appropriate treatment in the individual patient; it incorporated an *in-vivo* response to aromatase inhibitor treatment by measuring Ki-67 centrally at diagnosis and following a two-week pre-surgical exposure to either letrozole or anastrozole. The trial confirmed the low risk of recurrence for those with a low Ki-67 (>10%) when measured at diagnosis. In patients with a Ki-67 of >10% in the diagnostic biopsy, the value measured after starting pre-operative endocrine therapy provides additional clinical utility by dividing the cohort into those that respond to the endocrine therapy and show a drop in Ki-67 levels to below 10% and those that maintain their high proliferation levels. The former group have long-term outcomes similar to the intrinsically proliferation-low group, while those in the latter do much worse.

POETIC was a large, well-conducted trial undoubtedly providing sufficient evidence to enable us to raise the clinical utility status to 1-A on the Simon scale. Unfortunately, we are severely hampered in our abilities to recognize the full potential of that clinical utility due to problems of analytical reproducibility and transportability. This is true to the extent that the International Ki-67 in Breast Cancer Working Group (IKWG), in its very recent review of progress acknowledged the following key facts [10]:

(1) Ki-67 immunohistochemistry (IHC) has limited value for treatment decisions due to questionable analytical validity.
(2) Analytical validity of Ki-67 IHC can be reached with careful attention to pre-analytical issues and calibrated standardized visual scoring.
(3) Currently, clinical utility of Ki-67 IHC in breast cancer care remains limited to prognosis assessment in stage I/II breast cancer.
(4) Further development of automated scoring might help to overcome some current limitations.

We will consider all these points in greater detail in the subsequent sections of this review.

3. The Work of the IKWG

The IKWG came into existence in recognition of the central importance of Ki-67 in breast cancer prognosis and the lack of standardization of its analytics that was severely limiting its utility. It held its first workshop in 2011. At the meeting, the wide variability that existed in all areas of Ki-67 analysis became clear, and as a result the group planned a series of experiments to examine their sources.

3.1. *Phase 1: Visual assessment of a TMA using non-standardized methods* [11]

The first study used a 100-case tissue micro-array (TMA), from which centrally stained sections were distributed to eight self-designated "expert" laboratories for scoring using local in-house methodologies. The participating laboratories also received an unstained section which they stained and scored, again using local methods. Additionally, six laboratories undertook an exercise of repeat scoring the same stained array section on three consecutive days. Reproducibility was quantified by intraclass correlation coefficient (ICC), which produces a statistic that runs between 0 and 1, indicating no and perfect agreement, respectively. Between laboratories, reproducibility was moderate on centrally stained samples (ICC = 0.71, 95% CI = 0.47 to 0.78), and only borderline for local staining (ICC = 0.59, 95% CI = 0.37 to 0.68). When intra-laboratory reproducibility was examined, those using some form of counting outperformed laboratories using an estimation "eye-balling" method. Factors contributing to interlaboratory discordance included tumor region selection, subjective assessment of staining positivity scoring method, and counting method. In all probability, the potential for the first of these factors, tumor region selection, to introduce a significant degree of variability, is small in the TMA setting, and the question of what constitutes a positive nucleus is comparatively straightforward to deal with (see Figure 3).

3.2. *Phase 2: Visual scoring of a TMA using a standardized method* [12]

In this study, participating laboratories ($N = 16$) trained and calibrated their visual scoring using a training material set and a detailed counted scoring

Positive: **any** nuclear chromogen staining in an invasive breast cancer cell, above the surrounding background level.

Negative: **complete absence** of chromogen staining in the nucleus of an invasive breast cancer cell.

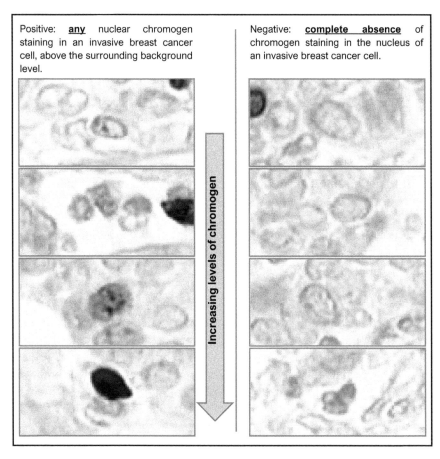

Figure 3: What counts as positive? These figures are taken from a web-based Ki-67 calibrator made available by the International Ki-67 in Breast Cancer Working Group as part of a set of training material to enable observers to have a reference set of Ki-67-positive and negative nucleus images. It is important to note any degree of specific chromogenic staining is sufficient to render a nucleus "positive". This figure shows an abbreviated image set, for a full-set please follow the web-link (http://www.gpec.ubc.ca:8080/tmadb-0.1/calibrator/index).

method provided by a web-based application. After this, they scored a 50-case TMA that had been centrally stained [12]. Training improved performance across the laboratories (mean Root Mean Square Error decreased from 0.6 to 0.4, maximum Absolute Deviation, from 1.6 to 0.9) and after completing this the ICC was 0.94 (95% Confidence

Interval (CI): 0.90–0.97), indicating the ability of laboratories to achieve very closely aligned scoring. However, questions still remained in regard to the applicability of this close agreement in the clinical setting using core biopsies and resection blocks stained in local laboratories.

3.3. *Phase 3: Visual scoring of core-biopsies using a standardized method* [13]

Sections from primary breast cancer core biopsies ($N = 30$) were centrally stained and the slide set was circulated among the 22 participating expert laboratories. Each laboratory scored Ki-67 using two methods:

(1) **Global** (four high-power microscope fields, with 100 cells evaluated in each).
(2) **Hot-spot** (single field from area of highest proliferation, in which 500 cells were evaluated).

A third method, the **weighted global**, was calculated from the global scores using weightings derived from estimated percentages of total area occupied by Ki-67-positive cells.

The ICC statistic was once again used to measure interlaboratory agreement, for the global method it was 0.87 (95% CI: 0.81–0.93), which met the pre-specified success criteria for reproducibility. The ICC for the hot-spot method was slightly inferior to that of the global (ICC = 0.84; 95% CI: 0.77–0.92), but more significantly, it failed to meet the pre-specified success criteria. While that for the weighted global (ICC = 0.87; 95% CI: 0.7999–0.93) indicated no improvement to justify the added complexity of its calculation.

The group had moved one step closer to its goal of defining a reproducible scoring method, but the superiority of global over hot-spot was by no means conclusive and lacked any clinical end-point corroboration.

Unfortunately, few independent comparative studies exist in the literature, but those that do speak in favor of the global score's superiority. In the study of Shui *et al.* published in 2015, a group of five observers assessed 160 breast cancer cases using the two methods and found an ICC for the global method of 0.904 compared to 0.894 for the hot-spot method, again there is only marginal superiority shown by the global method, and in common with the Leung *et al.* study there is no linkage to clinical outcome [14]. Very recently, there has been an advance in addressing the latter deficiency with the publication of a paper by Robertson and colleagues that used automated image analysis algorithms to produce hot-spot and global scores in a group of 139 whole-slide images of breast cancer sections stained for Ki-67 [15]. Results were assessed for their prognostic utility in predicting recurrence free survival (RFS) and overall survival (OS). Hot-spot was the superior predictor of RFS as defined by the hazard ratio (HR) for Ki-67-high versus Ki-67-low cohorts, HR: 6.88 (CI: 2.46–22.58, $p < 0.001$); in comparison, global scoring results produced an HR of 3.13 (CI: 1.41–6.96, $p = 0.005$). Conversely, global scores were the slightly better predictors of OS, producing an HR of 7.46 (CI: 2.46–22.58, $p < 0.001$), compared to 6.93 (CI: 1.61–29.91, $p = 0.009$) for hot-spot. Most significantly, multi-variate analysis demonstrated only global scores to be an independently significant predictor of both RFS and OS.

3.4. *Phase 4: Visual scoring of resection cases using a standardized method* [16]

In its final study in the area of scoring reproducibility, the IKWG looked at scoring in sections from excision blocks, and once again they compared global and hot-spot scoring. This time 23 observers took part, each scoring 30 centrally stained excision cases. The global method again showed superior assessor agreement as measured by the ICC among the group, which was 0.87 (95% CI: 0.799–0.93), marginally meeting the group's pre-specified success criteria of both ICC and CI being ≥0.8. The ICC for hot-spot was 0.83 (CI: 0.74–0.90), thus it failed to meet the success criteria in regard to the lower bound of the 95% CI.

Figure 4 shows a timeline for this work, and the parallel automation workstream illustrating the Group's progress in refining the process of scoring Ki-67 in a variety of clinically relevant scenarios.

The group has also produced further extremely robust evidence to support the need for standardization of the scoring method across the pathology community. In this study, which is the largest of its kind to be reported, the level of agreement across an arbitrary 20% cut-point was minimal (Cohen's kappa = 0.52). Considering no variability was introduced by pre-analytical or IHC methodological differences as all observers were scoring from an image of the same centrally stained 3 mm core,

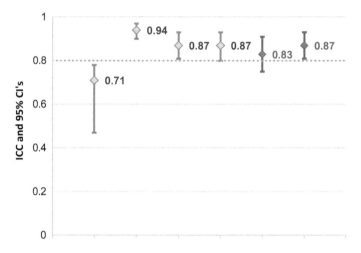

Figure 4:   International Working Group on Ki-67 in Breast Cancer's work is set-out in this time-line together with a pictorial representation of its progress, in the form of a plot illustrating the overall ICC (diamond boxes) and 95% Confidence intervals "whiskers" for each of its studies. Both manual and automated workstream are shown [3]. Key to references to the original papers shown in Refs. [4–9].

the result adds further support to the viewpoint expressed by Dowsett *et al.* writing almost a decade ago that: *Scoring procedures however vary at present, and their lack of standardization … is problematic. Thus, the direct application of specific cut-offs for decision making must be considered unreliable unless analyses are conducted in a highly experienced laboratory with its own reference data* [17]. It would seem that nothing has changed in the intervening years.

4. Pre-Analytical Factors

No one would doubt the impact variations in pre-analytical procedures, namely tissue acquisition, fixation, processing, and sectioning, can have on the reproducibility of Ki-67 scores. While many studies have investigated pre-analytics for Ki-67, no specific ASCO/CAP guidelines have been produced, as they have for the breast steroid hormone receptors (estrogen and progesterone receptors, ER and PR) [18] and the human epidermal receptor-2 (HER2) [19]. In the absence of specific guidelines, it would be good practice to adopt those published for ER, PR, and HER2 as the principles and mechanisms they are based on hold true for all antigen–antibody interactions. An outline is presented as follows, together with Ki-67 specific supporting evidence:

- *Shorten the time between tissue removal and fixation initiation to reduce ischaemic change artefacts.* In a series of experiments designed to test the resistance of Ki-67 to changes induced by fixation-delay, Arima and colleagues showed 10 hours' delay did not cause a perceptible change, but delays of 16 hours reduced mean scores by about 16% [20]. This would indicate a trend for little or no impact on core biopsies and only in cases where excisions were not incised and/or sliced before placing in fixative (formalin penetrates human tissues at a rate of about 1 mm per hour and the chemical reactions of fixation proceed comparatively slowly, taking up to 4 hours to complete [21]). Real-world data from the POETIC clinical trial support the postulated reduction of Ki-67 scores in resection specimens as there a relative 20% reduction in Ki-67 scores was observed on average when diagnostic core results were compared with matched resections in the no-treatment arm of the study [22]. And similarly changes in gene expression are seen according to sample type and duration of fixation delay [23].
- *Control the duration of fixation duration.* Essentially, all reports that have correlated duration of fixation with changes in protein expression levels as demonstrated by IHC have shown that extended duration leads to reduced protein levels being demonstrated. This is true regardless of the target antigen/antibody clone, and it is no different for Ki-67, at least in regard to the MIB1 clone. In their study on breast cancer resection cases, Arima *et al.* showed a relative fall in expression of 60% when optimal duration fixation (48 hours) was compared with 56 days ($p = 0.0003$). But, in practice it is probable that under-fixation is much more commonly encountered in the routine surgical pathology department increasingly under turn-around time pressure to quicken its processes. In the context of routine clinical laboratory practice, their observation that after very short duration fixation (three hours), expression levels were 60% lower relative to matched samples fixed for 48 hours is very worrying.

- *Control section drying conditions.* There is strong evidence that anti-genicity of cut-sections reduces over time, and that the rate at which this occurs depends on the specifics of the antigen involved and the conditions under which cut-sections are stored. Findings across multiple studies seem to suggest Ki-67 is comparatively robust, albeit taking into consideration that almost all the evidence has been obtained using the antibody MIB1 to demonstrate the protein [24–26]. As an example, the study by Andeen and colleagues used a TMA comprising four tumor types with two cores taken from each of 17 tumors in total (none were breast) [26]. Serial sections were stained for Ki-67 using the MIB1 antibody immediately after sectioning and at set timepoints thereafter, and the staining was evaluated by DIA in all cases. In the sections stained at six-weeks, the median H-score was 93% of that at time 0 when they had been kept at 4°C during the interim period and 89% for those stored at room temperature, both decreases are within acceptable limits of variation when taken in the context of overall staining variability. The trend for decreased reactivity being positively correlated with elapsed time continued out to the maximum examined in the paper of one year, by which time the relative reductions were 62 and 68%, respectively. Again, taking the results in the context of routine clinical practice, the deteriorations in Ki-67 staining are unlikely to significantly affect reproducibility. But this would not hold true for retrospective research studies using pre-cut sections.
- *FFPE block storage.* Much less evidence is available concerning the loss of antigenicity in stored FFPE blocks, but what there is suggests that decline in staining is not seen, or if it is present, then the phenomenon happens at a much reduced rate compared to cut sections [27, 28]. For example, one group has shown a relative decline in Ki-67 signal of 10% over 4.5 years as measured by quantitative immunofluorescence in TMA sections taken from more than 1,200 breast cancer cases [27].

5. Automated Scoring

The first reports on the use of automated digital image analysis (DIA) systems in the assessment of Ki-67 labeling began to appear in the mid-1980s, not very long after the FFPE-reactive MIB1 was described [29, 30]. On the whole these studies, while producing promising results showing good agreement with traditional manual scoring methodologies, tended to be *ad hoc* one-off projects looking at small numbers of cases on research orientated DIA machines, rather than concerted efforts at the implementation of field-changing workflows. And, while research interest in the subject remained constantly strong, it wasn't until the 2010s that things began to change in that regard.

Few commentators have doubted the ability of DIA to deliver highly reproducible results when pointed at the same image or set of images repeatedly. By their nature of operation, machines are inherently superior to any human in that respect. However, other obstacles have prevented the widespread uptake of DIA across the pathology community; principally these have been accessibility, implementation, and infrastructure. Until recently, image analysis software has required specialized "high-end" computing power to carry out the necessary analysis, and more importantly to store and manipulate the large image files generated by high-resolution scanning. The technology to deliver these same large images from remote servers in a timely fashion has also been lacking in non-specialist institutions and in routine diagnostic hospital departments in particular. Access to expensive scanning equipment has also been a limiting factor.

However, in recent years the information technology "landscape" has seen huge step changes driven by our increasing reliance on IT for the delivery of every aspect of our working and home environments. What were once regarded as "lightning-fast" processors are routinely found in desk-top personal computers or even laptops; high-capacity, high-speed networks, either cabled or wi-fi-based, are the norm in the office and at home, and we routinely view our work using large monitors at ultra-high definition and beyond. This, together with the availability of cheap storage that has more than kept up with all the other advances means that the age of digital pathology is truly upon us.

Even the DIA software is becoming more accessible and intuitive in its use. As an example, the APP's software solutions offered by Visiopharm A/S (Denmark, https://visiopharm.com/app-center/) tailored along the same lines as mobile phone applications.

6. Cognition Master Professional Suite

Another example of a simple to implement and easy to use DIA software system is the Cognition Master Professional Suite (VMscope GmbH

Berlin, Germany) [31]. The Ki-67 Quantifier application which is part of the Suite can be used on whole-slide images or captured images of a single high-power microscope field. It has no settings that can be altered, but is simply point-and-click, but its simplicity belies its ability to produce accurate, reproducible scoring results. In a recently conducted study of the Ki-67-stained slides from the POETIC clinical trial, which was mentioned earlier, the Ki-67 Quantifier was used to analyze core biopsies and resections from both arms of the trial, and a high correlation between manual and digital scores was observed in cores ($r = 0.92$, 95% CI = 0.87–0.94, $p < 0.0001$) and resections ($r = 0.95$, 95% CI = 0.86–0.98, $p < 0.0001$) and there was no significant bias between them ($p = 0.45$). This was a comparatively small study on only a small subset of the whole trial, but it certainly demonstrates the ability of DIA to produce comparable results to those given by a previously highly validated manual scoring system on trial materials with long-term clinical outcome data [32].

The first report on the use of the Cognition Master image analysis software came in 2015, from Klauschen *et al.*, who used it to re-examine the GeparTrio randomized clinical breast cancer trial [33–35].

Measurement of Ki-67 had previously been done centrally as part of the trial's translational research studies. The protocol for the manual scoring required capture of a single representative photomicrograph at a total magnification of x200 from each core biopsy using a standard microscope-camera setup and the evaluation and counting of 200 consecutive invasive tumor nuclei. In each case, the same photomicrograph was used for Cognition Master analysis without any post hoc image manipulations.

It is interesting to note that the reported number of images which could be counted manually was 1,116, while those suitable for DIA was 1,082. Therefore, there was an attrition rate of about 7% with the main reasons being sub-optimal pre-analytics leading to poor specimen quality, or poor imaging. This is useful information when thinking about how the system might "hold up" in a real-world setting, and indicates it is pretty robust. Most importantly, all statistical analyses performed demonstrated a high level of agreement between manual and DIA results; for example, the overall correlation (Pearson) was $r = 0.89$, and the ICC was 0.93. Hazard ratios for both disease-free survival and OS when calculated using the two scoring modalities were in very close agreement.

Although the study shows good evidence for the practical usability of this simple image analysis system, there are a few caveats that need to be kept in-mind. The central specimen handling, especially the IHC staining, would have led to a uniformity in the appearance of the preparations and therefore portability to other laboratories and reproducibility across laboratories cannot be inferred. And, while there are good overall agreements at the population level, there are still substantial numbers of cases at the individual level that would be differently classified across cut-points as can be inferred from the scatterplot presented in Figure 5. The scatterplot also shows a substantial bias in favor of higher manual scores been obtained when like-for-like cases are compared, indicating the need for recalibration of any cut-points derived by manual assessment if automated scoring were to be adopted. Very similar findings have been obtained in the author's own work with Cognition Master Ki-67 measurements; Figure 5 shows a set of results presented in a

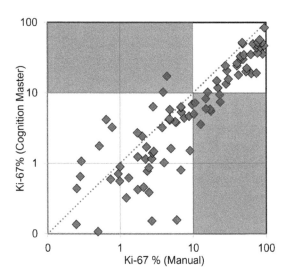

Figure 5: Scatterplot of manual versus automated scores for a subset of cases in the POETIC clinical trial [10]. The automated scores were generated by the Ki-67 application from the Cognition Master Ki-67 Image Analysis Suite. Red-shaded boxes enclose cases which disagree across an arbitarily chosen categorical cut-point of 10%. Despite an excellent correlation of 0.84 and an almost perfect ICC of 0.93 there are substantial numbers of cases falling inside the shaded boxes. Also of note is the bias in the results as shown by the greater numbers of points lying to the right of the line of equality (red-dotted line), which is indicative of a consistent tendency for the manual scores to be higher than their automated counterparts. Note the logrithmic (log_{10}) axes used to improve clarity around the 10% cut-off.

scatterplot that have an overall appearance almost identical to that shown by Klauschen *et al.*

7. Virtual Double Staining

When assessing Ki-67 labeling in breast cancer, or indeed any tumor type, it is important to be able to separate the invasive tumor component from non-invasive (in epithelial lesions, this is carcinoma *in situ*), and non-tumoral components. An experienced pathologist can do this almost without consciously considering it, using the cellular morphology and underlying intrinsic characteristics of the pathological "landscape". But, in some cases absolute certainty cannot be reached in regard to the invasive nature of a cellular cohort, in which case additional testing may be required.

An example of this is in the differentiation of *in situ* from invasive lesions by the use of p63 or similar markers of the myoepithelium, which is present in normal breast ducts and glands and also *in situ* disease but absent in invasive lesions.

Automated image analysis software does not have the advantage of years of training and experience when it is called on to analyze a field of cells for their intrinsic proliferation rate and therefore fully automated hands-free DIA systems suffer from the potential pitfall of averaging results across all cell types in a field. Careful selection of high-power microscope fields enriched for invasive cells can circumvent the problem, but for truly hands-free analysis, a method of automated invasive tumor "calling" is an essential requirement.

A solution to this problem has been explored by the Vyberg group in a series of studies utilizing a novel virtual double-staining (VDS) capability present in the Visiopharm software solutions [36, 37]. In their first paper, the group use a pan-cytokeratin (pan-CK) reactive antibody clone cocktail (AE1/AE3) to stain a series of 2 mm-core TMA blocks, where each core represented a sample of invasive breast tumor ($N = 140$ tumors in total), serial sections were also stained for Ki-67 using the MIB1 clone. Manual counting was undertaken for each case in the MIB1 stained TMAs, with a minimum of 200 cells being counted in each core. Scanned images from the pan-CK and MIB1 sections were aligned using the Visiopharm software to rotate and, where needed, local deformation. The pan-CK-positive areas were segmented as "tumor", all non-staining areas were classified as "stroma". The tumor-mask was then applied to the MIB1-stained matching core, allowing only nuclei within the masked area to be included in the scoring. In total, 103 cores were evaluable; correlation coefficient and ICC were in agreement, with both indicating almost perfect correlation between manual and DIA scores (r = 0.96; ICC = 0.98, 95% CI = 0.97–0.99). In keeping with almost all other comparative studies, there was a bias in favor of higher DIA scores compared to manual ones, but the trend was small (0.4% mean difference). The study has several limitations which need to be taken into account; it is a single institution study using highly optimized IHC staining protocols, the samples are TMA cores, albeit quite large ones, and the sampling area selected for coring was carefully chosen to exclude *in situ* disease, which would be a significant confounding factor in "real-life" samples, especially resections and wide-local excisions. Most significantly, an additional pan-CK IHC stain must be produced adding time and cost to the process. In its favor, the VDS procedure requires only standard IHC to be undertaken rather than using a more complicated and lengthy true double-staining method. The extra cost can be offset against savings in pathologists' time and the fact that standardized and reproducible automated counting translates into better patient outcomes.

In a further paper, they examined comparatively Ki-67 results obtained using VDS and DIA in 41 breast carcinomas stained using the six most commonly used antibody clones against MIB-1 and found significant differences in the mean Ki-67 score obtained using different antibody clones, different formats of those clones, and different platforms. When used in its ready-to-use (RTU) format, the primary antibody 30-9 (Ventana, USA) on the Ventana-supplied BenchMark ULTRA platform was significantly associated with higher positive cell counts compared to the mean (relative difference: 37.6%, $p < 0.001$), whereas the MIB-1 (Agilent Dako, USA) and the MM-1 (Leica Biosystems) clones, both in RTU format, produced staining of significantly fewer cells when used on the commercial suppliers own platforms (Autostainer and Bond, respectively), with the relative differences being −33.6% and −42.5%, respectively (both $p < 0.001$). Using concentrated antibodies titrated optimally on individual staining platforms largely (but not completely) negated relative staining differences, as exemplified by MIB-1, which had a non-significant relative difference in cells stained of 6.9% when used on the

Autostainer [38]. This well-conducted study highlights the large amount of variation that is introduced by analytical differences and the need to standardize the same if we hope to achieve reproducibility in the measurement of Ki-67 by IHC.

Further validation of VDS has been shown in a more recent paper from the same group in which they made comparisons of the VDS scores obtained for Ki-67 proliferation index in a TMA of fifteen invasive breast cancers, each represented by a single 3mm core, with the mean of those given by 199 observers using their preferred in-house method (which included estimation/"eyeballing", manual counting of between 100 and 1,000 cell nuclei per sample, and DIA). The group reported "excellent correlation between the two", but unfortunately did not report any statistical data to confirm this [39].

8. IKWG and Automated Ki-67 Scoring

David Rimm's group working out of Yale University have conducted parallel DIA studies on behalf of the IKWG, complementing those being done on manual scoring. Their first DIA study was an open-platform trial of scanners and analysis systems reported in 2019 [40]. It differed from the manual analysis project in that it required no pre-specified training or alignment. Fourteen laboratories each examined the same set of 30 pre-stained core biopsy samples as used by Leung *et al.* using their own in-house operating procedures. The labeling scores obtained were centrally evaluated for interlaboratory agreement using the ICC statistic. Categorical agreement was assessed using the kappa statistic and visually depicted in the construction of heat-maps. The ICC and 95% credible interval were similar for both studies, being 0.87 (95% CI: 0.81–0.93) for the manual and 0.83 (95% CI: 0.73–0.91) for the all-platform DIA study. Thus, demonstrating the potential for DIA to deliver reproducibility at least as good as that seen among trained expert manual observers.

In a second study that recapitulated the work done in both phases 3 and 4 of the visual scoring studies, Rimm's group settled on an open-source software product named QUPath (for its automated image analysis) [41, 42]. Programming guidelines were provided to 17 participating centers to enable them to develop their own analysis module for Ki-67 separately for the core biopsy images and the excision images (same 30-case sets used by the visual scoring group). Results were very similar to those obtained in the visual studies: ICC on core biopsies was 0.90 (95% CI: 0.85–0.95) and on excisions, 0.85 (95% CI: 0.78–0.91). As in the visual scoring, setting agreement is better on core biopsies compared to excisions, probably reflecting the wider scope for choice of scoring area in the latter setting.

9. Data from External Quality Assessment Schemes

External quality assessment (EQA) is the process of centrally assessing the results of a procedure submitted from multiple participating laboratories. Thus, participation in EQA allows laboratories to benchmark results of their own testing services with those of their peers both locally and worldwide.

The UK National External Quality Assessment Scheme for Immunocytochemistry and In-Situ Hybridization (UK NEQAS ICC & ISH) is one of the longest established examples of such EQA schemes with a global presence [43]. The Scheme does not have a Ki-67-specific proficiency testing program but has assessed the quality of staining for Ki-67 in breast cancer samples on a number of occasions and has published its findings [44]. Analyzing the data from 374 laboratories, submitting between them 2,601 slides, allowed them to identify substantial differences in the quality of staining with different antibody clones in general and when used on particular analytical platforms. Specifically, the clones 30-9 (Ventana, USA) and K2 (Leica Biosystems, USA) showed superior and equivalent performance when compared to MIB-1 (Agilent Dako). When it was used on the Autostainer platform (Agilent Dako), MIB-1 was significantly more likely to be associated with good staining ($p < 0.0001$) and its performance was similar to that of 30-9 and K2, whereas use on the Leica Biosystems Bond staining platform significantly reduced its likelihood of been associated with good staining ($p < 0.0001$). This nicely illustrates the ability of EQA schemes to assemble large data sets that reflect real-world testing, and in the case of UK NEQAS, globally derived data sets. The disadvantage is the disassociation from clinical outcome, and at this time, the lack of truly quantitative assessment methods, which is universally true across all EQA schemes.

The Nordic immunohistochemical Quality Control (NordiQC) Scheme is very similar in size and worldwide reach to UK NEQAS ICC & ISH [45]. It has also published on its experiences with Ki-67.

In its most recent paper, VDS and DIA were used to produce a set of gold-standard scores in 15 breast cancer cases represented as 3 mm cores in a TMA [39]. Results were compared to those obtained by nearly 200 NordiQC participants scoring digital images of the same TMA. The overall ICC was 0.85 (95% CIs not stated) and there was excellent agreement between DIA scores and the mean of the manually obtained scores.

Similarly, the Qualitätssicherungs-Initiative Pathologie (QuIP), a joint venture between the German Society of Pathology and the German Association of Pathologists, has conducted Ki-67 assessments in breast cancer specimens [46]. In their published study of 2017, the group reports on a series of six round-robin exercises conducted annually between 2010 and 2015. In these, unstained sections from a TMA containing cores from eight different breast cancer samples were circulated to up to 160 participating laboratories situated in Germany, and the surrounding countries. After local staining, observers were required to evaluate the proportion of positive stained nuclei in each core and ascribe it to one of four "bins", namely: ≤10%, >10–≤15%, >15–24%, ≥25%. Inter-observer concordance increased during the study period, with 86% of participants achieving excellent agreement (kappa >0.8) in the first years to 96% of participants achieving excellent results in 2015. The level of agreement was not dependent on methodological considerations and at the extremes of the spectrum, i.e., ≤10% and ≥25%, it was almost perfect by the end of the study. Those participants who remained in the study for multiple rounds improved their performance in proportion to the number of rounds undertaken, indicating the ability of EQA participation to improve concordance.

There is no doubt that EQA has a central role to play if Ki-67 is to establish itself as the biomarker it has so long promised to be. Future directions for EQA Schemes will be to provide laboratories with data to allow themselves to align more closely with one another, and to maintain that alignment, regular distributions using unstained sections of multiple samples need to be conducted (at intervals of no more than 4 months), most likely in the form of a TMA. Quantitative results from the local staining and scoring of the samples will need to be returned for central collation and production of anonymized consensus, which would be available to all laboratories allowing them to make comparisons with their own results. This will be critical for a value reported by any laboratory to be interpreted with confidence.

10. Positive Controls for Ki-67 IHC

It is accepted good laboratory practice to include at least a single positive control or, in many cases, multiple controls on the same microscope slide as the test material when conducting IHC staining. The subject would merit a book chapter all to itself and it is clearly beyond my remit to deal with it fully here, and therefore the reader is referred to an excellent review by Cheung and colleagues [47], part of the so-called "Evolution" series that deals with all aspects of quality assurance for IHC [48–50]. And also, to the prior report from the Immunohistochemistry Ad hoc Expert Committee's paper on the subject of controls [51], which introduced the concept of immunohistochemistry critical assay performance controls (iCAPC's). An iCAPC is a well-characterized tissue that has been shown to demonstrate predictable levels and patterns of expression of the antigen in question, and in which the cellular localization of specified different expression levels are well defined. In the case of IHC targets that are assessed in a semi-quantitative fashion such as Ki-67 and the majority of predictive biomarkers, the controls need to speak to the level of positivity, which is a question above and beyond the simple positive/negative question that is often addressed by diagnostic IHC.

For Ki-67, reactive lymphoid tissue such as tonsils removed for treatment of chronic inflammation is often the tissue of choice as it contains anatomical regions showing distinct, reproducible levels of proliferation, a fact long recognized as demonstrated by Gerdes *et al.* in their use of tonsil tissue to illustrate the staining of the Ki-67 antibody clone as described in their first paper on the subject (see Figure 1) [2]. Within the dark zone of the follicle center, which is usually orientated toward the bottom of the follicle furthest away from the point of entry, the proliferation index is maximal at approximately 60%. Elsewhere in the follicle center, the proliferation index is usually lower (at about 20–30%), and in the inter-follicular T-cell rich regions, much lower (at about 1%).

While the relative proliferative pattern of reactive lymphoid tissue described above remains generalizable from one lymph node or tonsil to the next, the absolute values of Ki-67 expression can still vary considerably, not only because of inherent biological variation but also because of variations induced by extraneous pre-analytical variation. When we attempt to align Ki-67 IHC methodologies as closely as possible, and once they are aligned, assure their continued alignment, an absolutely fixed and reproducible control or, even better, set of controls is very much more to be preferred. Cell lines are the "gold-standard" in such situations.

The obvious advantage in using cell lines as opposed to tissues is that individual samples are comparatively homogeneous if prepared correctly, they can be prepared in large volumes, and batch-to-batch variation is minimal. And, unsurprisingly the use of cell lines as controls for IHC goes back a long way, both in the history of the science, and as Ki-67 controls in particular. One of the first papers on it was published in 1996 by Ruby and McNally [52]. Their paper is not only ground-breaking in the use of cell line controls prepared using the same pre-analytical procedures to those used for routine histological samples, but also prototypical of almost all cell line controls that have been used subsequently right up until today in that the desired feature is an intrinsic biological characteristic. In this case, proliferation at a pre-characterized rate. The disadvantages of cell lines are also well-illustrated by the paper of Ruby and McNally in that the proliferation rates of the two chosen cell lines are about 50% and 90%, both much higher than the critical range we might be measuring in when using Ki-67 in the context of breast pathology. Here, cut points typically lie around the 5–20% range, as exemplified by the ≤13.25% (low) vs. >13.25% (high) cut-point proposed by Cheang *et al.* to discriminate luminal A from luminal B intrinsic subtypes [53], and the work of Tashima *et al.* that identified an optimal cut-off point of 20% as the most effective prognostic factor for Luminal/HER2-breast cancer [54]. Unfortunately, by their very nature of being able to survive in cell culture, cell lines are often highly proliferative, and those that do show low rates of proliferation are *ipso facto* slow growing and often problematic to maintain in long-term culture. In a survey of 17 commonly used breast cancer cell lines, Subik and colleagues found only two with proliferation rates of 30% or less, namely MCF-10A (a basal type cell line with a proliferation rate of 30%) and SKBR-3 (a HER2-positive cell line with a proliferation rate of 20%). The remainder all had proliferation rates of between 70 and 100% [55].

While it is true that individual cell lines do have characteristic rates at which they cycle, they are also biological systems and as such they are not absolutely locked-in to that rate, but rather they show some variations around that rate which may be condition or time dependent. Thus, batch-to-batch variation is present in cell line controls. A factor that has to be recognized and standardized against if true reproducibility of IHC is to be achieved.

The NordiQC EQA group has recently published on its experience with cell line controls [56]. The three cell lines used by the group were not specified other than their commercial source. Their expression levels were 70–80% and 80–90% (x2), respectively. Staining of the cell lines and three breast cancer tissue samples was conducted on three staining runs over the course of three separate days to examine reproducibility of staining on the platform used (BenchMark ULTRA supplied by Ventana). Results were assessed by DIA and reported as H-scores. For the cell lines, the coefficient of variation for all cores and days was <5% with the exception of a single outlier that represented a technical staining failure, indicating the feasibility of using cell lines in the EQA setting.

Recently, a new type of cell line control has been developed for Ki-67 IHC based on a concept proposed by Rimm *et al.* The new control is produced from two cell lines, each with known levels of expression, which are mixed together in varying regulated proportions thus allowing a set of controls to be produced each with a carefully tailored proliferation rate [57]. This innovative product has been further developed and commercialized by Array Sciences [58]. In its first produced format, the two cell lines were a human T-cell lymphoma derived line (Karpas) with a proliferation index of 80% and a caterpillar cell line (Sf9) in which the Ki-67 protein was sufficiently evolutionarily far from the human version so as to show no cross-reaction with any commercially available anti-human Ki-67 antibody. Validation studies have shown excellent reproducibility across the full range of tested proliferation rates, and in particular at the clinically relevant range lying between 5 and 20%. This control shows great promise in its ability to deliver a truly precision-tailored control for Ki-67 IHC.

11. Ki-67: Where Are We Now?

At the time of writing in early 2021, we stand at the cross-roads for the implementation of Ki-67 proliferation index scoring by the use of IHC as a fully validated clinical tool. Huge strides have been made in the establishment of its clinical utility in the field of ER-positive early breast cancer, particularly by the IKWG, but questions still remain with regard to its full potential as a companion diagnostic in other areas. In particular these include, but are not limited to, the following:

(1) Better determination of cut-points related to particular clinical outcomes, or perhaps better still, the elimination of reliance on single cut-points by the use of continuous measures.

(2) Application of validated automated scoring systems to increase precision in the measurement of the protein.
(3) Use of molecular techniques to measure *MKI67* gene expression as a replacement for Ki-67 measurement by IHC.
(4) Establishment of appropriate EQA programs.

In conclusion it may be appropriate to quote the authors of an editorial written as commentary on the consensus paper of the IKWG [10]. *Despite the remaining questions about Ki-67, it is gratifying to witness the results of collaborative, multi-disciplinary work and how they can provide objective, evidence-based answers to concerns that all practitioners must confront* [59].

References

[1] Scholzen, T., Gerlach, C., and Cattoretti, G. (2018). An insider's view on how Ki-67, the bright beacon of cell proliferation, became very popular. A tribute to Johannes Gerdes (1950–2016). *Histopathology* 73(2): 191–196.
[2] Gerdes, J. *et al.* (1983). Production of a mouse monoclonal antibody reactive with a human nuclear antigen associated with cell proliferation. *Int. J. Cancer* 31(1): 13–20.
[3] Gerdes, J. *et al.* (1984). Cell cycle analysis of a cell proliferation-associated human nuclear antigen defined by the monoclonal antibody Ki-67. *J. Immunol.* 133(4): 1710–1715.
[4] PubMed. *Ki-67.* [cited 2020 25/11/2020]; Available at https://pubmed.ncbi.nlm.nih.gov/?term=Ki-67&sort=date.
[5] Brown, D. C. and Gatter, K. C. (2002). Ki67 protein: The immaculate deception? *Histopathology* 40(1): 2–11.
[6] Cuylen, S. *et al.* (2016). Ki-67 acts as a biological surfactant to disperse mitotic chromosomes. *Nature* 535(7611): 308–312.
[7] Luporsi, E. *et al.* (2012). Ki-67: Level of evidence and methodological considerations for its role in the clinical management of breast cancer: Analytical and critical review. *Breast Cancer Res. Treat.* 132(3): 895–915.
[8] Simon, R. M., Paik, S. and Hayes, D. F. (2009). Use of archived specimens in evaluation of prognostic and predictive biomarkers. *J. Natl. Cancer Inst.* 101(21): 1446–1452.
[9] Smith, I. *et al.* (2020). Long-term outcome and prognostic value of Ki67 after perioperative endocrine therapy in postmenopausal women with hormone-sensitive early breast cancer (POETIC): An open-label, multicentre, parallel-group, randomised, phase 3 trial. *Lancet Oncol.* 21(11): 1443–1454.

[10] Nielsen, T. O. Leung, S. C. Y., Rimm, D. L., Dodson, A., Acs, B., Badve, S., Denkert, C., Ellis, M. J., Fineberg, S., Flowers, M., Kreipe, H. H., Laenkholm, A. V., Pan, H., Penault-Llorca, F. M., Polley, M. Y., Salgado, R., Smith, I. E., Sugie, T., Bartlett, J. M. S., McShane, L. M., Dowsett, M., Hayes, D. F. (July 1, 2021). Assessment of Ki67 in breast cancer: Updated recommendations from the international Ki67 in breast cancer working group. *J. Natl. Cancer Inst.* 113(7): 808–819, doi: 10.1093/jnci/djaa201. PMID: 33369635.

[11] Polley, M. Y. *et al.* (2013). An international Ki67 reproducibility study. *J. Natl. Cancer Inst.* 105(24): 1897–1906.

[12] Polley, M. Y. *et al.* (2015). An international study to increase concordance in Ki67 scoring. *Mod. Pathol.* 28(6): 778–786.

[13] Leung, S. C. Y. *et al.* (2016). Analytical validation of a standardized scoring protocol for Ki67: Phase 3 of an international multicenter collaboration. *NPJ Breast Cancer* 2: 16014.

[14] Shui, R. *et al.* (2015). An interobserver reproducibility analysis of Ki67 visual assessment in breast cancer. *PLoS One* 10(5): e0125131.

[15] Robertson, S. *et al.* (2020). Prognostic potential of automated Ki67 evaluation in breast cancer: Different hot spot definitions versus true global score. *Breast Cancer Res. Treat.* 183(1): 161–175.

[16] Leung, S. C. Y. *et al.* (2019). Analytical validation of a standardised scoring protocol for Ki67 immunohistochemistry on breast cancer excision whole sections: An international multicentre collaboration. *Histopathology* 75(2): 225–235.

[17] Dowsett, M. *et al.* (2011). Assessment of Ki67 in breast cancer: Recommendations from the international Ki67 in breast cancer working group. *J. Natl. Cancer Inst.* 103(22): 1656–1664.

[18] Allison, K. H. *et al.* (2020). Estrogen and progesterone receptor testing in breast cancer: American Society of Clinical Oncology/College of American Pathologists guideline update. *Arch. Pathol. Lab. Med.* 144(5): 545–563.

[19] Wolff, A. C. *et al.* (2018). HER2 testing in breast cancer: American Society of Clinical Oncology/College of American Pathologists Clinical Practice guideline focused update summary. *J. Oncol. Pract.* 14(7): 437–441.

[20] Arima, N. *et al.* (2016). The importance of tissue handling of surgically removed breast cancer for an accurate assessment of the Ki-67 index. *J. Clin. Pathol.* 69(3): 255–259.

[21] Fraenkel-Conrat, H. and Olcott, H. S. (1948). Reaction of formaldehyde with proteins; cross-linking of amino groups with phenol, imidazole, or indole groups. *J. Biol. Chem.* 174(3): 827–843.

[22] Smith, I. Robertson, J., Kilburn, L., Wilcox, M., Evans, A., Holcombe, C., Horgan, K., Kirwan, C., Mallon, E., Sibbering, M., Skene, A., Vidya, R., Cheang, M., Banerji, J., Morden, J., Sidhu, K., Dodson, A., Bliss, J. M.,

and Dowsett, M. (November 2020). Long-term outcome and prognostic value of Ki67 after perioperative endocrine therapy in postmenopausal women with hormone-sensitive early breast cancer (POETIC): An open-label, multicentre, parallel-group, randomised, phase 3 trial. *Lancet Oncol.* 21(11): 1443–1454, doi: 10.1016/S1470-2045(20)30458-7. Erratum in: Lancet Oncol. 2020 December; 21(12):e553. PMID: 33152284; PMCID: PMC7606901.

[23] Gao, Q. *et al.* (2018). Major impact of sampling methodology on gene expression in estrogen receptor-positive breast cancer. *JNCI Cancer Spectr.* 2(2): pky005.

[24] Jacobs, T. W. *et al.* (1996). Loss of tumor marker-immunostaining intensity on stored paraffin slides of breast cancer. *J. Natl. Cancer Inst.* 88(15): 1054–1059.

[25] DiVito, K. A. *et al.* (2004). Long-term preservation of antigenicity on tissue microarrays. *Lab. Invest.* 84(8): 1071–1078.

[26] Andeen, N. K. *et al.* (2017). Epitope preservation methods for tissue microarrays: Longitudinal prospective study. *Am. J. Clin. Pathol.* 148(5): 380–389.

[27] Combs, S. E. *et al.* (2016). Loss of antigenicity with tissue age in breast cancer. *Lab. Invest.* 96(3): 264–269.

[28] Ehinger, A. *et al.* (2018). Stability of oestrogen and progesterone receptor antigenicity in formalin-fixed paraffin-embedded breast cancer tissue over time. *APMIS* 126(9): 746–754.

[29] Franklin, W. A. *et al.* (1987). Quantitation of estrogen receptor content and Ki-67 staining in breast carcinoma by the microTICAS image analysis system. *Anal. Quant. Cytol. Histol.* 9(4): 279–286.

[30] Charpin, C. *et al.* (1987). [Microphotometric system of analysis with automatic scanning and immunodetection: Application to tissue sections (breast, endometrium, uterine cervix, ovary]. *Ann. Pathol.* 7(4–5): 285–296.

[31] Cognition Master. (2020). Cognition master professional suite. Available at http://vmscope.com/index.php?id=11.

[32] Alataki, A. Zabaglo, L., Tovey, H., Dodson, A., and Dowsett, M. (August 2021). A simple digital image analysis system for automated Ki67 assessment in primary breast cancer. *Histopathology* 79(2): 200–209, doi: 10.1111/his.14355. Epub 2021 May 5. PMID: 33590538.

[33] von Minckwitz, G. *et al.* (2008). Neoadjuvant vinorelbine-capecitabine versus docetaxel-doxorubicin-cyclophosphamide in early nonresponsive breast cancer: Phase III randomized GeparTrio trial. *J. Natl. Cancer Inst.* 100(8): 542–551.

[34] von Minckwitz, G. *et al.* (2008). Intensified neoadjuvant chemotherapy in early-responding breast cancer: Phase III randomized GeparTrio study. *J. Natl. Cancer Inst.* 100(8): 552–562.

[35] Klauschen, F. *et al.* (2015). Standardized Ki67 diagnostics using automated scoring — clinical validation in the GeparTrio breast cancer study. *Clin. Cancer Res.* 21(16): 3651–3657.

[36] Roge, R. *et al.* (2016). Proliferation assessment in breast carcinomas using digital image analysis based on virtual Ki67/cytokeratin double staining. *Breast Cancer Res. Treat.* 158(1): 11–19.

[37] Lykkegaard Andersen, N. *et al.* (2018). Virtual double staining: A digital approach to immunohistochemical quantification of estrogen receptor protein in breast carcinoma specimens. *Appl. Immunohistochem. Mol. Morphol.* 26(9): 620–626.

[38] Roge, R. *et al.* (2019). Impact of primary antibody clone, format, and stainer platform on Ki67 proliferation indices in breast carcinomas. *Appl. Immunohistochem. Mol. Morphol.* 27(10): 732–739.

[39] Røge, R. Nielsen, S., Riber-Hansen, R., and Vyberg, M. (February 1, 2021). Ki-67 proliferation index in breast cancer as a function of assessment method: A NordiQC experience. *Appl. Immunohistochem. Mol. Morphol.* 29(2): 99–104, doi: 10.1097/PAI.0000000000000846. PMID: 32168036.

[40] Rimm, D. L. *et al.* (2019). An international multicenter study to evaluate reproducibility of automated scoring for assessment of Ki67 in breast cancer. *Mod. Pathol.* 32(1): 59–69.

[41] Acs, B. Pelekanou, V., Bai, Y., Martinez-Morilla, S., Toki, M., Leung, S. C. Y., Nielsen, T. O., and Rimm, D. L. (January 2019). Ki67 reproducibility using digital image analysis: An inter-platform and inter-operator study. *Lab Invest.* 99(1): 107–117, doi: 10.1038/s41374-018-0123-7. Epub 2018 Sep 4. PMID: 30181553.

[42] Bankhead, P. *et al.* (2017). QuPath: Open source software for digital pathology image analysis. *Sci. Rep.* 7(1): 16878.

[43] *UK National External Quality Assessment Scheme for Immunocyto-chemistry and In-Situ Hybridisation.* (2020). Available at https://ukneqasiccish.org/.

[44] Parry, S., Dowsett, and M., Dodson, A. (February 1, 2021). UK NEQAS ICC & ISH Ki-67 data reveal differences in performance of primary antibody clones. *Appl. Immunohistochem. Mol. Morphol.* 29(2): 86–94, doi: 10.1097/PAI.0000000000000899. PMID: 33337635; PMCID: PMC7993918.

[45] Vyberg, M. and Nielsen, S. (2016). Proficiency testing in immunohisto-chemistry — experiences from Nordic Immunohistochemical Quality Control (NordiQC). *Virchows Arch.* 468(1): 19–29.

[46] QUiP. (2020). *Qualitätssicherungs-Initiative Pathologie.* Available at http://www.quality-in-pathology.com/.

[47] Cheung, C. C. *et al.* (2017). Evolution of quality assurance for clinical immunohistochemistry in the Era of precision medicine: Part 4: Tissue tools for quality assurance in immunohistochemistry. *Appl. Immunohistochem. Mol. Morphol.* 25(4): 227–230.

[48] Cheung, C. C. *et al.* (2017). Evolution of quality assurance for clinical immunohistochemistry in the Era of precision medicine: Part 1: Fit-for-purpose

approach to classification of clinical immunohistochemistry biomarkers. *Appl. Immunohistochem. Mol. Morphol.* 25(1): 4–11.

[49] Torlakovic, E. E. *et al.* (2017). Evolution of quality assurance for clinical immunohistochemistry in the Era of precision medicine. Part 3: Technical validation of immunohistochemistry (IHC) assays in clinical IHC laboratories. *Appl. Immunohistochem. Mol. Morphol.* 25(3): 151–159.

[50] Torlakovic, E. E. *et al.* (2017). Evolution of quality assurance for clinical immunohistochemistry in the Era of precision medicine — part 2: Immunohistochemistry test performance characteristics. *Appl. Immunohistochem. Mol. Morphol.* 25(2): 79–85.

[51] Torlakovic, E. E. *et al.* (2015). Standardization of positive controls in diagnostic immunohistochemistry: Recommendations from the international ad hoc expert committee. *Appl. Immunohistochem. Mol. Morphol.* 23(1): 1–18.

[52] Ruby, S. G. and McNally, A. C. (1996). Quality control of proliferation marker (MIB-1) in image analysis systems utilizing cell culture-based control materials. *Am. J. Clin. Pathol.* 106(5): 634–639.

[53] Cheang, M. C. *et al.* (2009). Ki67 index, HER2 status, and prognosis of patients with luminal B breast cancer. *J. Natl. Cancer Inst.* 101(10): 736–750.

[54] Tashima, R. *et al.* (2015). Evaluation of an optimal cut-off point for the Ki-67 index as a prognostic factor in primary breast cancer: A retrospective study. *PLoS One* 10(7): e0119565.

[55] Subik, K. *et al.* (2010). The expression patterns of ER, PR, HER2, CK5/6, EGFR, Ki-67 and AR by immunohistochemical analysis in breast cancer cell lines. *Breast Cancer (Auckl)* 4: 35–41.

[56] Røge, R. Nielsen, S., Riber-Hansen, R., and Vyberg, M. (February 1, 2021). Image analyses assessed cell lines as potential performance controls of Ki-67 immunostained slides. *Appl. Immunohistochem. Mol. Morphol.* 29(2): 95–98, doi: 10.1097/PAI.0000000000000845. PMID: 32168035.

[57] Aung, T. N. Acs, B., Warrell, J., Bai, Y., Gaule, P., Martinez-Morilla, S., Vathiotis, I., Shafi, S., Moutafi, M., Gerstein, M., Freiberg, B., Fulton, and R., Rimm, D. L. (July 2021). A new tool for technical standardization of the Ki67 immunohistochemical assay. *Mod. Pathol.* 34(7): 1261–1270, doi: 10.1038/s41379-021-00745-6. Epub 2021 Feb 3. PMID: 33536573; PMCID: PMC8222064.

[58] LLC, A. S. (2021). *Array Science.* [cited 2021 02/01/2021]. Available at http://www.arrayscience.com/.

[59] Reis-Filho, J. S. and Davidson, N. E. (July 1, 2021). Ki67 Assessment in breast cancer: Are we there yet? *J. Natl. Cancer Inst.* 113(7): 797–798, doi: 10.1093/jnci/djaa202. PMID: 33369665; PMCID: PMC8246841.

https://doi.org/10.1142/9781800611399_0006

Chapter 6

Multiplex: From Acquisition to Analysis

Regan Baird* and David Mason†

Visiopharm A/S, Agern Allé 24, 2970 Hørsholm, Denmark

*trb@visiopharm.com
†dma@visiopharm.com

1. Introduction

1.1. *Why multiplex?*

Histological tissue specimens contain thousands to millions of individual cells captured and preserved in place at a specific moment in time of disease progression. A pathologist can recognize hallmarks of normal and abnormal tissue to make qualitative and semi-quantitative assessments. But knowing the genes and proteins expressed by each cell has the potential to unlock a deeper understanding of the state of the disease, leading not only to more personalized treatments, but also identification of responders to therapy. Pathologists may visually identify the lineage of a cell based on its morphological and staining appearance but cannot measure the level of gene expression, presence of surface proteins, or the cells' activation state. Dissolution of the tissue would individualize the cells for single-cell analysis using PCR, flow cytometry, or other analytical methods, but information about cellular interactions and neighborhoods would be lost. Ideally, a pathologist would like to get access to spatially resolved and multi-dimensional information from the tissue, reflecting genetic and protein biomarker expression levels within each cell. Histological multiplexing, the labeling of a tissue for multiple biomarkers with high specificity, will allow further definition of the phenotype of each cell labeled in the tissue. With multi- to high-plexed staining techniques and increasingly sophisticated imaging technology, there is a concomitant and growing demand for software tools enabling a detailed and spatially resolved analysis of such high-dimensional image datasets. This chapter will attempt to describe state of the art for such analysis tools, as well as point to various trends and unmet needs for this type of analysis.

1.2. What is phenotyping?

While multiplexing techniques provide markers that indicate the presence or absence of specific features, typically proteins, they cannot themselves provide cell identity. This requires the translation of specific markers into a cell's identity in a process called phenotyping.

When thinking about the tumor microenvironment, one of the key considerations is the role of the immune system in regulating, controlling, and, in some cases, exacerbating the disease state. Despite the complexity of the markers involved, many hallmark cellular epitopes that can be used to phenotype cells of the immune system have been identified. A simple example is shown in Figure 1 where three different populations of T cells can be identified based upon the three cell markers CD4, CD8, and FoxP3.

For any given experiment, assuming that cells are considered either positive or negative, the theoretical maximum number of phenotypes that

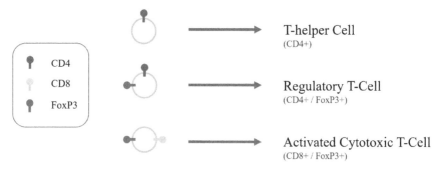

Figure 1:   A simple example of phenotyping three cell types based on the presence (or absence) of three phenotypic markers.

can be classified is 2^n, where n is the number of markers. Thus, in the example above where $n = 3$, eight different phenotypes can be identified (2^3). These permutations represent both the presence and absence of markers. It should be noted that the *lack* of a marker can be as relevant, and indeed as phenotypically crucial, as its presence.

When looking at all conceivable permutations, it is important to keep in mind that not all combinations of probes will be biologically relevant. A simple example mentioned above, is a triple positive CD4+, CD8+, FoxP3+ cell, which is likely to be extremely rare [17]. This topic is covered in greater detail in Section 2.2.

1.3. *Chromogenic multiplex*

1.3.1. *Chromogenic staining*

Traditional Pathology requires a highly trained expert to "read" the tissue on a slide through different powers of a microscope. The tissue is prepared by staining it to enhance the contrast of different structures in the slice of the tissue. Haematoxylin and Eosin (H&E) and other special stains differentially absorb within the tissue to reveal to the pathologist the state of the disease. Pathologists are trained and experienced in visually interpreting all the hallmarks of a disease in the patient's sample and making a proper diagnostic assessment. In the 1980s, immunohistochemistry (IHC) techniques were developed and introduced to the clinical pathologist. IHC uses antibodies targeted to specific proteins expressed in cells along with a counterstain (typically haematoxylin, staining the nuclei in light blue) to identify cells that do or do not express that antigen. This will give a true positive and negative count for any biomarker tested. The positive cells will stain depending on the chromogen used, typically in dark brown (chromogen DAB (3,3'-Diaminobenzidine)) or red (chromogen Fast Red), both of which contrast well with the counterstain light blue (Haematoxylin), allowing for easy distinction.

Classic examples are the hormone receptors in breast cancer, where individual adjacent tissue sections can each be stained with one of ER, PR, HER2, or Ki67 antibodies. When immunostaining is carried out correctly, the color absorption of the cell is a function of the level of biomarker expression in each cell. This panel of four individual biomarkers has been used extensively by pathologists to more properly assess the progress of breast cancer, design personalized treatment regimens, and develop targeted pharmaceuticals specific to a patient's expression level/ combinations of these biomarkers. One challenge with this panel is that it is on four individual slides rather than being four biomarkers on a single slide. While there are historical and technical reasons for this, it would be more convenient and require less tissue if these markers were combined on a single slide and were distinguishable from each other when colocalized to the same cell. It is also necessary for the pathologist to identify invasive tumor areas to properly assess each biomarker. Adding the biomarker p63 to the panel would clearly separate invasive and preinvasive tumor areas, thereby increasing the accuracy of the measurement. Understanding the level of immune infiltrate into the tumor through immune cell biomarkers introduces another set of markers which are of increased interest in several indications. Unfortunately, the swatch of chromogen dyes available to stain tissue is limited; these dyes can bleed into and compete on the tissue; and as the dyes start to overlap, they mix into a dark color that can be hard to distinguish. Typically, two or three chromogenic dyes per tissue slice is an optimal balance of information density and accurate analysis, however, multiplexing of 4–5 biomarkers with chromogenic dyes has also been successfully proven to be a possibility.

1.3.2. *Virtual multiplex*

Virtual multiplexing is a technique that co-registers two or more digitized images of stained tissue sections, considering translation, rotation, and local nonlinear deformations that typically occur between FFPE tissue sections. This is one approach to overcoming the limitations of chromogenic multiplexing described in the previous section and to achieving high(er)-plexed chromogenic datasets. There are several advantages of this approach worth mentioning: First, pathologists are more used to interpreting chromogenic slides, which visually preserve the anatomical context. Moreover, there is currently no simple way to test for staining sufficiency in multiplexed datasets, and there is reason to believe that good staining quality/sufficiency is harder to achieve for high(er)-plexed datasets. With serial sections, where it is possible to assess stain quality/ sufficiency in each, critical aspects of data quality can be controlled and ensured. There are, however, also limitations of this approach. In particular, sections must be as close in space as possible, 3–5 microns, for the most accurate alignment, since the three-dimensional tissue structure will differ increasingly with growing distance. In addition, specific cells only appear in a few of the sections, which can make cellular co-expression analyses difficult.

There have been several practical applications of the virtual multiplexing approach, demonstrating its practical feasibility in both diagnostics and research.

1.3.2.1. Tumor and region identification

The compartmentalization of tissue into different regions of analysis is required for many assays. The evaluation of the aforementioned breast cancer panel biomarkers is always limited to the invasive tumor regions only. Virtual multiplexing is a proven tool to accurately identify these different areas prior to proper assessment of the respective biomarkers. Compartmentalization is a necessary precursor in many tissue assays, but it can be quite difficult to accurately discriminate and measure those areas visually if the staining combination necessary to sequester these areas and measure biomarkers is not compatible. This task can be better served by separating this in adjacent serial sections where each is stained under the ideal conditions for either finding the necessary regions or necessary biomarkers. Then the individual sections can be merged virtually using digital alignment. Specific staining of tissue areas by IHC will increase the precision of region identification by eliminating doubt and increase efficiency through automatic segmentation of these regions using easily developed artificial intelligence (AI). When non-IHC stains (e.g., H&E, Trichrome, Giemsa, etc.) are the only contrasting methods to identify the compartments, virtual multiplexing will also work between this stain and the biomarker IHC sections. The overall change in the volume of a tissue compartment such as tumor across adjacent (3 micron) serial sections is minimal [23]. Only the edges of the identified region should contain misalignments and that error can be accounted for in virtual multiplex assays[2].

1.3.2.2. High-plex colocalization studies

When engaging in high(er)-plex studies, it is important to understanding uses and limitations of virtual multiplexing. Initial studies of this method were undertaken by the Ref. [23]. As another example of high-plex applications, investigators at Moffitt Cancer Center were interested in determining the proximity of eight biomarkers with respect to tumors in lung cancer patients [15]. IHC was applied to continuous serial sections, one for each biomarker where the center section (i.e., Section 5) was stained for the tumor. Virtual multiplexing was performed with the nine sections (eight biomarkers + tumor) such that the tumor compartments could be properly demarcated on all biomarker sections to determine the localization of each biomarker in and around the tumor area. However, the cellular co-expression of the eight biomarkers was not feasible given the assay required several microns of tissue sections. We designed a small study to see if it was possible to detect cellular co-expression across aligned serial sections [11]. We attempted to determine if the same CD68 stained monocyte could be identified across three (4-micron) serial sections of tonsil. After co-registration of the sections, we found 60% of the CD68 signal overlapped between adjacent sections and 30% of the signal overlapped across all three sections, suggesting that we were able to detect the same monocytes in different sections. However, we also measured a 15% discontinuous overlap of the CD68 signal in Sections 1 and 3 but absent in the middle section, suggesting signal was from different but nearby monocytes. We concluded that it is not always possible to predict if the overlapping signal across serial sections is from the same cell spanning serial sections or two nearby cells expressing the same signal. The assumption would be even more difficult if the cell type were smaller than monocytes or if nuclear biomarkers were used. Yet, with careful assay design, virtual multiplexing across serial sections can be an empowering tool to leverage information from individually stained slides.

1.3.3. *Multiplexed immunohistochemical consecutive staining on single slide*

New methods to increasing the IHC multiplex on the same piece of tissue have been developed, including, e.g., an assay system designed at Mt. Sinai Hospital called *Multiplexed Immunohistochemical Consecutive Staining On Single Slide* (MICSSS) [1]. The principle of cyclic staining methods is to re-stain tissue after washing in repeated IHC cycles. In theory, this applies also for an unlimited number of biomarkers in a panel, because each independent round provides a highly interpretable staining of the same cells. The virtual image generated will contain a true co-expression of the biomarkers such that the phenotypic profile of each cell can be determined. Each staining round is adjusted to the ideal staining conditions for the biomarker without compromise. The tissue is then digitized followed by complete biomarker removal through gentle washing so it can be stained repeatedly. The digitized tissues are then co-registered and interpreted. Co-registration can be done one of two ways: the whole tissue registration is used also in virtual multiplexing, or by careful alignment of each nuclei in the images. The latter method works well because you can assume the nuclei are in the same location after each round as they should be from the same cell. Each tile of the WSI per staining round can be individually aligned at the nuclear level and then the WSI is restitched or rebuilt. The challenge with this cyclical staining method is to be very careful during the washing steps so as to not disturb the tissue, remove nuclei, or washout antigen. If the tissue does deteriorate over many rounds of washing and re-staining, then tissue co-registration by virtual multiplexing can be used for an improved alignment.

1.4. *Immunofluorescent multiplex*

1.4.1. *Immunofluorescence (IF)*

IF is becoming an increasingly popular tissue imaging technique. IF contrasting method works in reverse to chromogenic dyes because the positive signals are bright and the negative signals are dark. Unlike its chromogenic counterpart where overlapping dyes blend together, the detection of each fluorescent signal is distinct, and you can measure the contribution of each biomarker to a cell's total signal. Most IF instruments allow 4–5 biomarkers to be imaged simultaneously without interference (bleed through or crosstalk) and the images appear very crisp with high sensitivity given the detection range, resolution, and contrast to background. Cyclic IF methods are the fluorescent counterpart to MICSSS and include sequential immunostaining [5] and CYCIF [18] or are performed by commercial instruments like Akoya's CODEX, GE's Cell Dive, and Miltenyi Biotec's MACSima, to name a few, each using multiple rounds of staining on the same piece of tissue. Reagent manufacturers are starting to provide services and staining kits around cyclic methods, including Ultivue, NeoGenomics, and CellIDX. Ultivue's InSituPlex® technology uses proprietary antibody-DNA barcoding and signal amplification to enable same-day multiplex biomarker detection in tissue. There are different ways to remove the signal from a previous round, including washing

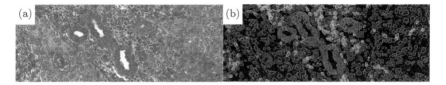

Figure 2: Virtual multiplex of H&E (a) and IF (b). CD3+ T cells (green), PCK+ Tumor Cells (red), and all other cells (blue) were identified in the IF image and transferred to an adjacent H&E image after virtual multiplex. Annotations on the H&E were then used to train a deep learning AI to find these cell types.

or photobleaching. These cyclic methods also require alignments of each tile or the whole-slide image as described above.

One complaint about IF images is that the non-stained areas are blank, lacking the contextual/anatomical information that chromogenic images provide where these areas are still illuminated and visible. By combining the chromogenic image with IF images through virtual multiplexing, it is possible to retain the contextual information. This provides the best of both imaging modalities, the context of the brightfield with the specificity of IF. This dual modality virtual multiplex is also very useful to automate the development of training annotations for deep learning (DL) algorithms. Figure 2 illustrates the power of the combination: an H&E image was co-registered with a serial section stained for epithelial tumors with cytokeratin and CD3 for T cells (among other stains from Ultivue). The annotation labels can be transferred to every tumor and every T cell in the H&E using the IF as a guide and become a new training set for AI. This trained AI could then recognize lymphocytes, T cells, and tumor infiltrating lymphocytes (TILs) in H&E images without the need for the corresponding IF image.

1.4.2. *Spectral unmixing*

IF instruments create images by capturing all the light emitted from fluorophores after selective excitation. But each emission spectra of single fluorochromes can be quite broad, and traditional fluorescent microscope instruments have difficulty distinguishing fluorochromes with overlapping spectra. Akoya has developed a technology that mathematically separates overlapping emission spectra from up to nine fluorochromes, allowing the largest simultaneous IF biomarker panel to date [22]. This spectral

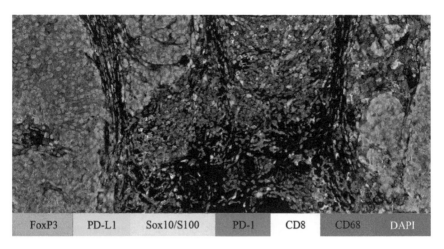

| FoxP3 | PD-L1 | Sox10/S100 | PD-1 | CD8 | CD68 | DAPI |

Figure 3:   Example of a spectrally unmixed multiplex seven-channel immunofluorescent image from Akoya Biosciences.

unmixing technology in Akoya's Vectra and Vectra Polaris instruments also separates the frequently overlooked autofluorescence signal that confounds tissue imaging into a separate channel (Autofluorescence is more than a mere contaminant as it can provide important information or be used as a signal during image analysis). The image can be quite picturesque and loaded with data from up to nine biomarkers (Figure 3 shows seven biomarkers). Analyzing all the phenotypic combinations in a 9-plex is quite challenging and requires computational assistance discussed in what follows. The 9-plex images can also participate in virtual multiplexing to further increase the numbers of biomarkers in a study.

1.5. *Imaging mass cytometry multiplex*

While cyclic techniques can theoretically be used to probe an unlimited number of antigens, practical factors such as antigen degradation, steric hindrance of antibodies, and tissue disruption caused by repeated rounds of staining can limit its utility. The need for multiple rounds of staining comes primarily from the spectral overlap in the detection probes, be they chromogenic or fluorescent. Several companies offer reagents for low-plex staining with both modalities which can push detection to, e.g., up to 4 chromogenic stains and 8 fluorescent stains in a single pass. Fluorescent probes, of course, require not only careful selection of reagents, but optical hardware (such as mirrors and filters) in the detector to be able to separate individual channels.

Mass cytometry imaging (MCI) uses the same concept for primary immunolabeling but radically re-thinks the approach to secondary antibody labeling and detection [3]. Instead of enzymes or fluorophores, these techniques use heavy-metal conjugated antibodies and *Time Of Flight* (TOF) mass cytometry to record the abundance of each isotope [9, 16], which corresponds to the amount of antigen present in that pixel. Unlike the potential of optical techniques, the lateral resolution is relatively low, being rarely below 300 nm. Reduced spatial information in the image can have some subtle impact for the pathologist, leading to harder manual interpretation and more difficult quality control. It can also impact automated analysis, given a higher overall background signal and less information with which to segment objects. These downsides (along with a considerably longer acquisition time) are offset however, by the ability to simultaneously label and acquire upwards of 35 labels, enabling a highly fine-grained phenotyping and cellular analysis in a single section.

2. Approaches to Phenotyping Using AI

2.1. *Cell segmentation*

Outlining cells in tissue sections is quite challenging for image analysis today, especially when the multiplex contains many overlapping cells and signals. Early high-plex images were analyzed by measuring biomarker signal overlap within pixels without pre-defining the cells that occupy the space [7, 8]. Identifying cellular boundaries is difficult because of the heterogeny of cellular morphologies in a tissue especially in late disease progression. Most people envision images of the distinct outer membrane of confocal microscopy images of a confluent cell monolayer, but this is not the reality in tissue. Tissue slices have a greater thickness than a monolayer and contain overlapping cells or parts of cells caught during the physical resection of the tissue. Multiple cells can therefore share the same pixel, thus sharing signal, and stains designed for segmentation are rare and generally not included in biomarker panels. Outer membrane biomarkers tend to have a high background in tissue because the membrane spans the cell and not just the 2D perimeter. All these contaminating signals will confound the segmentation of cells in the tissue.

2.1.1. *Machine learning approaches* (*classical*)

Most cell segmentation strategies start with defining the nucleus of every cell, because nuclear stains are typically a part of every multiplex panel. However, cells without a nucleus or a well-defined nucleus and multi-nucleated cells must be identified using an alternative strategy. Nuclear segmentation routines typically involve identifying the nuclear signal and applying roundness and size filters. The simplest approach to characterization of sub-cellular compartments with respect to, e.g., bio-marker response, is to expand the nuclear mask by a designated distance and assume that the cytoplasm and membrane are included. Adjacent and overlapping nuclei require additional rules of separation. This estimate of the cell boundary improperly assumes that the nucleus is in the center of the cell and its shape mirrors that of the rest of the cell. However, this method usually does an acceptable job of modeling cells in tissue.

2.1.2. *DL/AI* (*NextGen*)

DL-based classifiers have led to significant improvements in nuclear segmentation especially with tissues that have a large heterogeneity of nuclear sizes and densities. The DL reduces the number of concessions that older rule-based methods required, allowing for nuclear segmentation of various shapes and sizes. Nuclear segmentation will become commonplace, but this only identifies the nuclear boundary and not that of the entire cell. There are only two choices for defining the cell boundary: stain for it or estimate its location. In the absence of stains specific for cell segmentation, the boundaries can be defined by concatenating signals from other biomarkers in the panel known to localize to the cytoplasm or membrane. Be cautioned, though, that as more biomarkers are combined, the signals will widen and begin to overlap adjacent cells. The widened signal can be linearized to then separate cells, however, that filter may not always correctly bisect adjacent cells. Still, sophisticated algorithms can be engineered to take advantage of combining the DL-based nuclear segmentation approach with a few other signals from the panel to craft a routine that models all cells in the tissue sample within an acceptable error. Once every cell has been segmented, we can extract the location and signal intensity of all biomarkers (expression level) for every cell in the tissue.

2.2. *Semi-automated approaches to phenotyping*

Once a phenotypic "unit", such as a cell, has been defined, the question becomes, how to classify the cell as positive for any individual marker. The simplest way to do this is to use a threshold classifier. A cell can thus be classified as positive or negative based on an intensity cut-off.

In Figure 4, the intensity of a fluorescent channel is used to classify, however, brightfield imaging is also amenable to this approach. Brightfield images are almost always captured on an RGB color camera. Unlike fluorescent markers whose component intensity is usually contained within a discreet channel, chromogenic stains will have varying components across the Red, Green, and Blue components of the detector. If the color vector (that is the contribution to each of the RGB channels) is known, image analysis software can be used to unmix the input into a color vector matching the stains used (Figure 5).

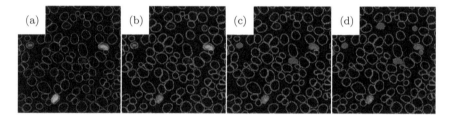

Figure 4:   Example of a threshold classifier used to differentiate negative (gray outlines) and positive (magenta filled) cells. Lowering the cut-off (a–d) classifies more cells as positive.

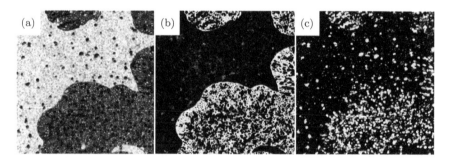

Figure 5:   Before using threshold cut-offs, Brightfield images need to be unmixed into a color vector representing a specific stain. (a) Original DAB/AEC stained image, (b) the DAB color vector, (c) AEC color vector.

In this way, any cell, be it from brightfield, fluorescence, or any other modality, can be classified according to its positivity.

For markers that are mutually exclusive, this approach is often enough. Single cells can be phenotyped and simple metrics can be reported, such as the number of cells and their positions. More likely however, subpopulations of phenotypes are required to add granularity. An example matrix that follows shows how even the presence of four markers allows for greater flexibility around phenotyping. Despite this, the enormous complexity of cellular immunology means that interpretation will always be limited by the granularity available. For example, a CD3 positive cell, while identifiable as a T cell, cannot be further classified without also including a secondary lineage marker (CD4/CD8) or functional marker (e.g., Granzyme B/Caspase 3).

In practical terms, scaling up the thresholding approach described above is simply a case of constructing a decision tree for each combination of markers. Given that each marker can be in one of two states (positive or negative), the number of possible combinations of markers (including negative for all markers) can be expressed as 2^n, where n is the number of markers. The four markers in Table 1, therefore, allow for 16 (or 2^4) combinations, but require 15 separate applications of an intensity cut-off (albeit one threshold per channel).

In cases where the biological classes of interest are already known, the complexity of this semi-automated approach can be reduced by only looking for classes of interest. For well-understood phenotypes (e.g., double-positive CD3/CD8 or CD3/CD4), this may be acceptable,

Table 1: An example matrix showing how the presence or absence of particular markers can be used to infer cellular phenotypes.

Phenotype	CD3	Granzyme B	Ki67	Cytokeratin
T cells	•	—	—	—
Cytotoxic T cells	•	•	—	—
Proliferating T cells	•	—	•	—
Proliferating cytotoxic T cells	•	•	•	—
Proliferating non-tumor cells	—	—	•	—
Tumor cells	—	—	—	•
Proliferating tumor cells	—	—	•	•

however, the presence or absence of novel or transient markers may not be as well understood. Likewise, the exploration of novel phenotypes may be the whole point of the investigation. In these cases, a semi-automated approach is not only laborious but can also lead to the introduction of bias around phenotype identification and selection.

There are several other reasons why a semi-automated approach to multiplex phenotyping is generally not advisable. One of the main limitations is the requirement of *a priori* knowledge of the intensity distribution of each channel. The simple but necessary inclusion of a range of probes inevitably incorporates variability, as a function of quantum efficiency and absorption. Combined with possible background signal from tissue structures, a simple threshold quickly ceases to perform with adequate sensitivity. Finally, this approach can lead to inconsistencies when considering markers with multi-modal intensity distributions, such as a biologically relevant weak positive, intermediate positive, and strong positive expression profile.

2.3. *Unbiased autophenotyping using AI autoclustering*

As the number of biomarkers in the panel increases, the number of combinations exponentially increases to the point that the manual construction of the logic tree for phenotyping becomes overwhelming. In flow cytometry, AI clustering is used for unbiased analysis of individual cells, and the population of cellular phenotypes is determined. This approach can be generalized to the analysis of spatially resolved and multi-dimensional tissue image datasets. Such autoclustering techniques vary, but in short, the algorithm uses AI to determine the number of overlapping biomarker object clusters, or phenotypes, based on the biomarker signal from a representative training set of segmented cells. The algorithm determines the level of signal necessary for positivity of each biomarker and renders a phenotype for each cell. It is also possible to include degrees of positivity in autoclustering approaches. This will, however, significantly increase the complexity of the data. Applying this analytical technique to intact tissue will provide additional morphological and positional data compared to flow cytometry but is accompanied by additional challenges.

Autoclustering relies on high-quality signal for each biomarker. Any artifacts, background, inconsistent staining, or improper tissue preparation will contribute to errors in the unbiased phenotyping. It is also at the mercy of the cell segmentation because the cell boundaries limit the area used to determine biomarker positivity. This means that if a signal inappropriately crosses a cell boundary, it would be included incorrectly in a cell's phenotype. Likewise, if a signal is outside the cellular boundary, it will not contribute to the cell's phenotype. If a small cell is adjacent to a bigger cell and the segmentation boundary is not modeled perfectly, it is possible for signals to cross the boundary causing the autoclustering to assign incorrect phenotypes.

The average signal intensity of each biomarker in each cell will distribute into three populations: positive, negative, and weakly positive cells. The challenge is finding the line between the negative and the weakly positive populations.

Autoclustering will unbiasedly determine a cut-off intensity value from the population of cells used to train the AI classifier. Limiting the mean intensity calculation of each cell to only the pixels in which the biomarker resides will increase the sensitivity of the autoclustering algorithm to determine positivity. For example, a nucleus may comprise less than half of the area inside a cell's segmented border (Figure 6), and the non-labeled pixels in the remainder of the cell will dilute the intensity measurement of nuclear biomarkers. Reducing the number of dim pixels will boost the signal of weakly positive cells while minimally influencing the average intensity of negative and strongly positive cells. Further, biomarkers that localize to the membrane or have a punctated expression pattern may require a further reduction in the number of pixels used for calculating biomarker positivity. Noisy signals or signals that inappropriately traverse a segmentation border can be remedied by removing the brightest pixels from the mean calculation, as they should not be associated with the cell.

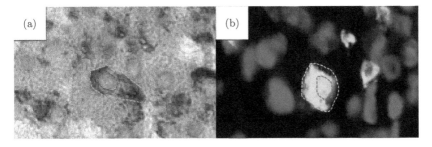

Figure 6:   Two examples of segmented cells showing the size of the nucleus (outlined in blue) and the entire cell (outlined in green). The nuclear compartment in each cell is approximately 25% of the total cell area. Note the difference in cell shape.

Autoclustering can be susceptible to false negatives when the dynamic range of signals is large because the weakly positive cells more resemble negatives than strongly positives in the distribution histogram. This is overcome by setting a positivity cap or overflow bin in the histogram, artificially decreasing the dynamic range, bringing the populations closer together, and thereby magnifying the weakly positive cells. It may also be useful to set a minimum cap to eliminate any noisy signals or high background from interfering with the autoclustering determination.

Once the mean signal intensity conditions are determined for each biomarker, the autoclustering routine can associate positivity of each biomarker for every cell. Low yield phenotypes are generally false, so it may be prudent to set a minimum count required to register as a phenotype yet remain wary to not eliminate true rare events. When there are many cells to phenotype, it is sometimes best to iterate the analysis by first setting the abundance threshold high, say 1 in 500 cells (0.2%) to register as a phenotypic class. This will reduce the number of phenotypes requiring verification and leave the remaining cells for the second iteration of low abundance clustering, say 1 in 10,000 limit (0.01%). Phenotypes deemed biologically irrelevant can be removed or reassigned to their appropriate class before quantification.

Another approach is to sort cells based on lineage markers first and then apply AI autoclustering using only the lineage-specific biomarkers from the panel. For example, first sort cells into lymphocytes, macrophages, tumor, and other cells if your panel contains the lineage markers like CD3, CD68, and Pan-cytokeratin, for example. Once sorted, each parent class can be examined individually for the remaining biomarkers of interest individually. This tactic should reduce the number of false phenotypes that need to be scrutinized but may prevent the discovery of new phenotypes.

3. Meaningful Data Extraction

In the following, we will discuss two broad approaches to interrogating a dataset. In the first example, the user has a specific question. Phenotyping is here used to identify a sub-class of cells to study. A second approach is a discovery or exploratory approach whereby the aim is to explore a tissue or region of interest and seek a broad understanding of the cell types represented.

3.1. *Targeted approach*

Image analysis of single probes in simple systems such as cultured cells is typically a question-driven endeavor. A metric such as positive percentage might be calculated for a given probe, or the distribution of probe intensities might be measured and classified. A significant problem exists when trying to extend this approach to tissue sections, given their heterogeneity. Depending upon the source organ, a section may contain stroma, parenchyma, and native or invasive immune cells. These are expected to be found in healthy tissue, but disease states further add to the complexity, with the addition of disease specific regions such as tumor, fibrosis, and necrosis. The problem is that if you are trying to measure the response or presence of a particular marker such as PD-L1, knowledge of the specific population in which a measurement is being made is important.

By means of an example, a study may seek to understand the immune response to solid tumor and thus want to determine what percentage of T cells within 20 μm of a tumor are cytotoxic. Measuring the presence of Granzyme B (a marker of cytotoxicity) across all cells provides an incorrect answer. Both cells outside the proximal region, and cells other than T cells may be Granzyme B positive. Thus, a targeted approach to identify a subset of cells within an area is required.

There are myriad ways to approach this problem. In the following, one possible workflow is presented, involving several steps. In our example, each step is implemented as an independent analysis solution, here: a Visiopharm APP.[1] The analysis begins as described in Section 2.1, with an APP for cell segmentation in the whole tissue sample using a pre-trained AI classifier (Figure 7 panels (a) and (b)). Here, the use of DL is key as the cells show significant variance in size, shape, intensity, and morphology. It is important to detect all cells as we do not know *a priori* which cells will be required for analysis.

In this context, an APP incorporates both a classifier (which can be simple, shallow, or DL-based) as well as an unlimited number of post-processing steps. These have wide-ranging utility from subclassifying objects (e.g., selecting based upon size or shape) to somehow altering existing labels previously created by a classifier (e.g., dilating cell masks to incorporate a peri-nuclear region or filling holes in a label).

---

[1]Analysis Protocol Packages or APPs are Visiopharm's modular image analysis "building blocks" that allow easy modification reuse and distribution of workflows.

Taking advantage of these post-processing tools we can easily change the label class based on SOX10 positivity (Figure 7 panel (c)) and proximity (Figure 7 panel (d)). This demonstrates how a subset of cells can be selected, although any combination of probe positivity, proximity, location, shape, or many other features can be used to subset cells of interest.

At this point, the cells of interest are used (as in Section 2.3) for multiplex phenotyping. Of note, these cells can also be used for the unsupervised training of a classifier, but typically this is done on the whole image or selected regions to incorporate all relevant phenotypes. Once the cells have been classified (see Figure 7 panel (e)), any output can be reported. We can now calculate the number of cells that are T cells (including basic T cells as well as those that are cytotoxic and proliferating) then calculate the fraction of these that are cytotoxic (which, in this case, means positive for Granzyme B).

This basic workflow highlights how some simple pre-processing ahead of multiplex phenotyping can allow users to ask very specific

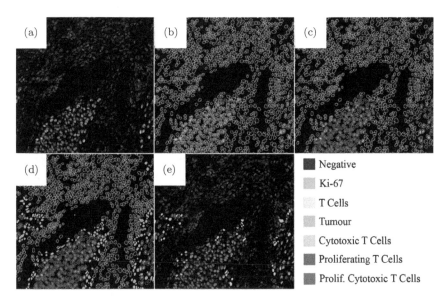

Figure 7: An example workflow used to answer a specific question, namely what percentage of T cells within 20 μm of tumor are cytotoxic. See the accompanying text for more details.

questions of their data while still taking into account the complexity of cellular phenotypes.

3.2. *Discovery approach*

A sequence of algorithms can be logically arranged in an automated work-flow. Each computational task in the sequence may employ different AI technologies to engineer the most precise and efficient algorithm. Tissue compartments, for example, can be manually annotated, thresholding based on biomarker intensity (e.g., cytokeratin for epithelial tumors), or by training an AI algorithm to identify morphological landmarks in the image.

(1) Co-register images if necessary (virtual multiplex).
(2) Run quality control algorithms to ensure proper staining and to remove artifacts from the analysis.
(3) Isolate the primary tissue area to avoid processing unnecessary portions of the glass slide.
(4) Compartmentalize tissue area into regions such as Tumor, Tumor Margin, Necrosis, Stroma.
(5) Segment every cell within the required regions.
(6) Automatic phenotyping using AI autoclustering finds all biomarker combinations in the region.
(7) Quantification of regions and cells.

Automatic phenotyping can feel a little bit like a black box allowing the AI to determine the phenotypes with little to no supervision. It is critical that the dataset used for the training of the Machine Learning classifier contains representatives of all possible phenotypes, including positive and negative examples. If the training set contained cells from only a normal control population that is then applied to a patient population, you would be able to determine differences in normal phenotypes but would miss phenotypes present in only the experimental population. Visual inspection of the images after autoclustering is a common practice, but be cautioned that image displays can be deceiving, especially when inspecting signals near the noise floor in fluorescent and IMC images. Be sure that the image color display is consistent and not adjusting to the maximum and minimum signals as you zoom, pan, or switch images. But if you find phenotyping errors from the unbiased autoclustering, it may be hard to understand what adjustments need to be made.

Data visualization tables and plots can be very useful in determining the quality of the unsupervised analysis and troubleshooting phenotyping errors by the autoclustering algorithm (Table 2). These visualizations include tables with the name and population count of each phenotype, phenographs showing the biomarker expression level heat map of each phenotype, cell lineage phylogeny maps, and two-dimensional reduction plots. The list of generated phenotypes may reveal some biologically unknown phenotypes, perhaps a new type of cell has just been discovered! More likely, though, one or more of the biomarkers has been misassigned to a phenotype. If it is detected that an unexpectedly high number of cells are negative for all biomarkers, the signals could be weak and the autoclustering may require higher sensitivity. If a high number of phenotypes deemed "unknown" or "not classified" is found, perhaps the number of allowable phenotypes has been exceeded, or the abundance criteria of phenotypes was set too strictly.

Multiplex images are considered high-dimensional data, thus reducing this information into simpler 2D visualizations using mathematical embedding methods makes the phenotyping data much more interpretable. 2D reduction plotting methods are invaluable tools for determining the reliability of the unsupervised autoclustering and troubleshooting those biomarker signals that need supervised intervention. t-distributed stochastic neighbor embedding (t-SNE, Figure 8) is an embedding technique that plots the phenotyping data such that objects with similar signals are close to each other, and objects with very different signals are further apart [20]. In multiplex, cells that express the same or similar biomarkers will be nearby, forming phenotypic clusters that can be colored by their density or biomarker expression levels. The original implementation of the t-SNE is limited in the number of data points it handles and was optimized for larger cytometry data sets under the proprietary vi-SNE [2]. Since 2008, there have been many other permutations of SNE. Another issue with t-SNE is that the spatial relationship of the clusters (the gaps between the clusters) is not preserved and plotting the data can take a long time. Uniform Manifold Approximation and Projection (UMAP) is a newer dimensional reduction technique that decreases the plotting time while preserving the spatial relationship between the phenotypic clusters [4]. The mathematics behind the reduction methods is complex and requires machine learning, but interpreting the plots is quite easy.

Table 2: Troubleshooting guide for automatic phenotyping using autoclustering algorithms and data visualizations.

Troubleshooting guide	Possible remedy	Alternative remedy
High negative cell count	Increase sensitivity by decreasing the dynamic range of the intensities	Reduce the number of pixels in the mean calculation to the highest intensity
Too many phenotypes	Limit the number of cells required to register as a phenotype	Limit the cell count requirement to a percentage of the total population
Too many undetermined phenotypes	Raise the number of cells required to register as a phenotype	Raise the cell count requirement to a percentage of the total population
False positive phenotypes	Increase the lower intensity limit for the mean intensity	Include more dim pixels when calculating the mean intensity
False negative phenotypes	Reduce the overflow value for the upper intensity limit for the mean intensity	Reduce the number of pixels to the brightest pixels when calculating the mean intensity
Signal crosses boundary and assigned to the wrong cell	Reduce the number of bright pixels when calculating the mean intensity	Modify the cell segmentation parameters
Biologically irrelevant phenotypes	Eliminate or reassign phenotype to the correct class	Adjust the autoclustering parameters for the incorrect biomarker

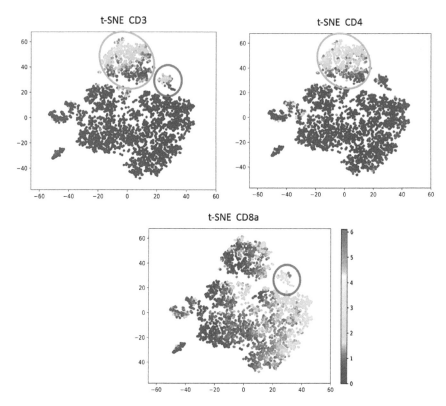

Figure 8:   t-SNE analysis of T cell biomarkers CD3, CD4, and CD8. The distribution of CD3 follows two populations, CD3 positive cells in the green circle cluster with CD4 positive cells and are the T-helper cell phenotype. The red circle encircles the CD3 and CD8 positive population and are the Cytotoxic T cells. Notice the high level of background in the CD8 (non-purple objects) suggesting that the sensitivity for CD8 was set too low.

For purposes of this discussion, we will use t-SNE plots. t-SNE plots are a redistribution of the autoclustering data where each point in the plot references a cell in the population and its color designation describes its phenotypic class or its biomarker expression level. Comparing the biomarker expression pattern of two or more t-SNE plots will reveal biomarkers that migrate together and biomarkers that do not. Biomarkers that have overlapping heat map on their t-SNE will be colocalized in the cells of the image, whereas biomarkers that do not overlap are not colocalized in cells. Figure 8 shows the t-SNEs of T cell biomarkers CD3, CD4, and CD8. Within each circle you can see T cells with varying expression levels of each biomarker shown by the heat map color scale, red objects highly express the biomarker down to purple objects representing cells that do not express the biomarker. Since CD3 is the lineage marker, we expect all T cells to express this biomarker and it clusters into two distinct populations (Green circle and red circle). The population in the green circle also expresses CD4, the T-helper cell population. The CD3 positive cell cluster in the red circle co-expresses CD8, the cytotoxic T cells. Notice that the CD4 and CD8 positive clusters do not overlap as expected because CD4 and CD8 co-expression on T cells is not known to happen biologically. Also notice the low level of expression of CD8 in many CD3-negative cells, which is generally not to be expected biologically (sometimes NK-cells can). It suggests the sensitivity for CD8 needs to be reduced in a supervised modification of the autoclustering procedure.

If the t-SNE plot of a biomarker shows an unexpected high level of background rather than a tight cluster area, increase the lower threshold settings for that biomarker. On the other hand, if only a few objects have a high signal, then lower the sensitivity for that biomarker. An unexpectedly wide and unchanging color distribution of a biomarker often suggests that it is a noisy signal or was poorly stained; if so, it probably should be eliminated from analysis. Overlapping hot spots of biomarkers not known to colocalize suggests that the segmentation of the cells is incorrect and may need to be adjusted — but first confirm that it was not a new cell phenotype that was just discovered! It is important to have biomarker controls in the panel to best assess the results of autophenotyping.

Careful interpretation of t-SNE plots can quickly illuminate information about the biomarkers beyond troubleshooting. As an example, IMC images of 41 breast cancer patients that responded to trastuzumab treatment and 19 patients that had a recurrence of disease after trastuzumab treatment were analyzed for 18 biomarkers [8]. We examined this data post-publication, and it was immediately clear from the t-SNE plots that the phenotype population cluster expressing the intracellular and extracellular domain of HER2, the trastuzumab target, was present in the responder group but missing in the recurrence group, confirming a known finding. We also noticed different phenotypic clusters between the t-SNE plots of each group for signal transduction pathway biomarkers in tumor cells (unpublished data), suggesting the tumor cells in responders were signaling differently than the patients experiencing recurrence. The t-SNE plots of lymphocyte markers revealed little differences in the clustering pattern of CD8+ cytotoxic T cells in the two patient groups, but there was a population of CD8-negative T cells that appeared co-expressed with tumor signals. The co-expression was likely artifacts of segmentation or the autoclustering misassigning neighboring signals, as it is unlikely that the T cells express tumor markers. However, these artifacts revealed a population of tumor-associated T cells easily visualized in the t-SNEs. Proximity analysis confirmed these T cells were significantly closer to cytokeratin positive tumor cells than other CD3+ populations (6 μm vs. 54 μm). Most interestingly, this population was captured in the responder cohort and not the recurrent patients, suggesting an increased immune response in these breast cancer patients after trastuzumab treatment. This cursory discovery project revealed many prospective targets that will be confirmed by orthogonal techniques.

Each phenotype is described primarily by the biomarker expression pattern, is the cell positive or negative for each biomarker? The expression pattern can be represented by a heat map as shown in Figure 9 where biomarkers with a high expression pattern in a given phenotype are colored in shades of red, low expression, in shades of blue, and intermediate expression, in pink/white/lavender. This phenotype matrix will show the expression level of each signal assigned to each phenotype and the count of cells in that phenotypic class. This plot can help determine if negative cells are truly negative and how the single-, double-, triple-, etc., positive phenotypes' expression patterns relate to each other. The autoclustering performed in Figure 9 identified some biologically nonrelevant phenotypes, for example, the CD4+ only population is unlikely real as is the CD4+Tumor and CD3+CD4+CD8 populations, since CD4 is typically associated with CD3 and, in all but some rare cases, exclusive from CD8. The CD4 expression signal in these nonrelevant phenotypes is a lavender blue color or weakly positive, suggesting the CD4 sensitivity was too high during autoclustering. In fact, the CD4 signal in the image appeared to have a high background and the t-SNE of CD4 had a wide but weak expression pattern (data not shown). So, the autoclustering is finding false positives due to the biomarker background and can be remedied by lowering the sensitivity or increasing the lower signal limit used during autoclustering. The other CD4 containing classes appear to have a high CD4 signal and are correctly assigned. Alternatively, the CD4 only cells could be simply reassigned to the negative category, the CD4+Tumor reassigned to the tumor only class, and the CD3+CD4+CD8 reassigned to the CD3+CD8 (cytotoxic T cell) class, as retuning the autoclustering could have additional unforeseen consequences. Similar inferences using

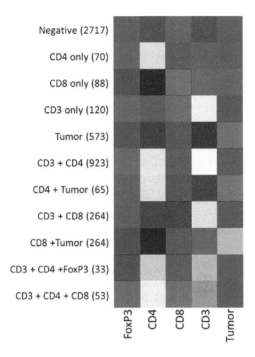

Figure 9:   Phenotype Matrix: Expression pattern of each phenotype in a six-plex IF study (including DAPI, FoxP3, CD4, CD8, CD3, and a tumor marker). Each row represents a phenotype identified by autoclustering with the cell count for that biomarker in parentheses. Each column represents the biomarker. Shades of red represent high expression levels and shades of blue represent low expression levels.

this matrix and t-SNE plots can be made for the other phenotypic classes in question. Similarly, examine the prevalence of phenotypes to identify over- and underrepresented phenotypic classes in each tissue slide or across study groups.

The amount of data compiled upon completion of an unsupervised autophenotyping analysis of a single piece of tissue can be large. Best practice is to create an iterative analysis strategy to begin first to appreciate the complexity of the image data and to digest the results of a preliminary round of autophenotyping. Then modify settings for only the biomarkers that are problematic rather than attempting to predetermine the autoclustering settings. Remember, the autoclustering AI is detecting signals from imperfect biological samples that are imperfectly stained and require biological expertise to make the necessary interventions for the best results. Once autoclustering results are confirmed, extract the cell count of each phenotypic class in each tissue compartment, the biomarker expression level of each biomarker in every cell, each cell's location, the tissue compartment it is found in, and its proximity to other phenotypes. The count of each phenotype is not necessarily the count of lineage biomarker. For example, to measure the total lymphocyte count, you must sum all the phenotypes that contain the parental T cell biomarker (e.g., CD3). As the number of biomarkers in the panel used in the T cell lineage increases, so does the complexity of the phylogeny. The phylogenetic analysis between compartments and then between study groups will also become proportionately more challenging. Consider analyzing each parental cell lineage separately to simplify the task. Querying other metrics such as size, shape, intensity, etc., of each cell can also help to understand the morphological effect of biomarker expression on cells and possibly be an added characteristic of the phenotype (e.g., a phenotype could be CD3+CD8+ cells between 5 and 10 microns in diameter and within 30 microns of tumor cells). Be aware that the data generated from multiple tissue slides in multiple study groups provides an opportunity to make many discoveries but can also lead to analysis paralysis.

Combining high parameter multiplex techniques with virtual multiplex can be a very powerful discovery tool as it simultaneously increases the number of biomarkers or structures that are identifiable and confirms findings through orthogonal measurements. Today's tissue co-registration technology allows for combinations of chromogenic (e.g., H&E) stained sections, multichannel IF-stained sections, and high parameter Imaging Mass Cytometry-stained sections to be registered down to a sub-cellular alignment to be analyzed as a single virtual image. The combination will provide all the benefits of each technique while overcoming each imaging modality's limitation. For example, the contextual appearance of the tissue and tissue compartmentalization could be provided by analysis of the chromogenic image, cell segmentation specific signals could be introduced using the high contrast and resolution of IF imaging, and IMC could increase the number of biomarkers required for higher order phenotyping. If one or more biomarkers is duplicated between the IF and IMC panel, then the assay will have an internal control to ensure good tissue section alignment and biomarker colocalization within cells. An added advantage of this multiplex is that when certain cell types are confirmed by the AI, they will be automatically segmented and annotated in all tissue sections of the virtual image. Thus, the cells would be annotated in the

H&E/IHC image and could be used in the training set for novel AI algorithms.

4. The Future of Multiplex

4.1. *Technological advances*

4.1.1. *Introduction*

The approaches and techniques described in the previous chapters represent an overview of breakthroughs, challenges, and gaps in an important, rapidly growing, but still emerging field. In our opinion, the ability to analyze spatially resolved and high-dimensional datasets in a robust way will propel tissue pathology into the realm of precision medicine, and become a valuable platform both for the development of new drugs and companion diagnostic biomarkers. The following paragraphs aim to highlight some recent or potential developments that may already or will impact the field of multiplex phenotyping.

4.1.2. *Probes*

As discussed above, one of the primary drivers of novel techniques around multiplex experiments is the inability to spectrally separate multiple probes with similar spectral profiles. In a field already decades old, novel fluorophores are still being developed as optimal chemistries, and techniques for measuring them are being improved [14]. These efforts are largely based on improvements to Quantum Efficiency or brightness, but when considering the multiplexing potential, the width of the emission spectrum is also an important factor. The thinner the emission band of the fluorophore, the more spectra can be fit into the sensitivity range of the detector.

In response to this challenge, several approaches have turned to nanocrystals or "Quantum dots", which have exceptionally narrow emission bands. These approaches can yield versatile and tuneable emission profiles but are not without problems of steric hindrance [10] and temporal blinking [13]. That development continues on in the hope of further increasing the potential of this approach.

Regardless of probe choice, the techniques described above all rely on immunolabeling of protein antigens to phenotype cells. Proteins are, of course, only one step in the functional description of a cell. *In situ* hybridization (ISH) techniques use complimentary RNA or DNA probes to label gene products within tissues and cells [12] and indeed these techniques in their modern incarnation form the basis of certain diagnostic tests [28].

A similar hybridization technique forms the basis of a relatively new technology that can be used to spatially resolving mRNA sequences in a tissue sample. Such *spatial transcriptomics* use DNA barcoding techniques to localize captured mRNA sequences during sequencing [27]. While the spatial resolution of commercial instantiations of these approaches is significantly lower than even Mass Cytometry Imaging techniques, the sheer number of different possible mRNAs that can be attributed to these regions makes this a technique with extraordinary potential for some types of multiplex phenotyping [21], and commercial systems with resolutions relevant for pathology are under development.

4.1.3. *Hardware*

Multiplex images can be acquired on almost any acquisition hardware. Small field of view (FOV) acquisition devices such as microscopes can acquire images quickly but without larger spatial context. Within the field of histopathology, whole-slide scanners are typically favored for their ability to capture a whole tissue or biopsy section. This allows not only for a larger phenotyping dataset but also the spatial context of where phenotypes occur with regard to tumor margins, tissue boundaries, and blood vessels. Whole-Slide Imaging has become a mature technology with several fast and high-quality devices capable of imaging up to 8 or 10 channels now commercially available. There is a certain level of specialization between (1) scanners for diagnostic purposes, with an *in vitro* diagnostic (IVD) approval, that are usually optimized for chromogenic images, and (2) scanners for research, with very high magnification and the ability to acquire multi-channel fluorescent images. Gradual improvements to the speed and accuracy of motorized stages and the sensitivity and data-throughput of acquisition cameras will continue to increase the throughput and quality of these techniques.

Assuming the same acquisition hardware is used, further enhancements to speed and utility can be seen upstream at the staining level. Cutting edge approaches with microfluidics are making serial-staining more automated, quicker, and more economical with staining reagents.

Another advantage of microfluidic systems is that they are small enough to be mounted in or on an acquisition device. This removes the need for the sample to be moved in between serial stains, improving the accuracy of the resultant multiplex, even if software methods are still used to register a stack of images.

As described earlier, Mass Cytometry Imaging techniques bypass some of the traditional limitations of optical probe systems. Their main limitations to broad adoption, however, are the acquisition speed and, perhaps as a result, limited acquisition FOV. This is an important point given the relevance of context in many multiplex analyses. Thus, advances in these exciting techniques are likely to focus on incorporation of more probes but an increase in speed perhaps by parallelizing sample collection.

4.1.4. *Software*

The complexity of high-plex image datasets is prohibitive to manual analysis. Software for deep and comprehensive analysis of highly multiplexed data are less mature and have only become commercially available over the last 5 years but are required to fully leverage the possibilities multiplexed imaging technologies offer. This is one of the areas where we expect the most development over the next 5–10 years.

4.1.4.1. Region segmentation with DL

One of the quantum leaps in the field has been the use of DL to assist in several aspects of the workflow. Given the structural complexity of both healthy and diseased tissue, it can be helpful to break down a section into its constituent regions. In immuno-oncology, this might mean specifically detecting tumor or invasive margins, which can provide another layer of granularity on top of phenotyping. Furthermore, if certain compartments or artifacts can be ignored, then processing time can be decreased. DL can also enable a very nuanced segmentation to improve output accuracy. Note in Figure 10 that thresholding techniques are not able to differentiate necrosis and tumor, both of which stain positive for cytokeratin. DL can be trained to recognize these compartments, allowing exclusion of necrotic areas. Of course, this is one example, and the same idea can be applied to excluding artifacts (such as tissue folds and tears), detecting blood vessels, or finding other structural landmarks.

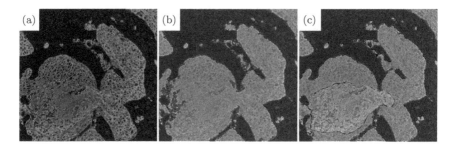

Figure 10: Tissue segmentation with DL. A tumor stained with Cytokeratin (a) contains a necrotic core. Thresholding techniques (b) cannot differentiate tumor from necrosis here, while DL (c) can successfully make the distinction.

The logical extension of this is to start to detect these features or structures using other non-specific channels or probes which nonetheless contain identifying information.

4.1.4.2. Cell detection with DL

We have discussed above semantic segmentation of tissue; however, DL is also making big impacts on cell detection, a critical part of the multiplex phenotyping workflow. Traditional techniques use either intensity-based thresholding (with or without additional morphological filters) or shallow learning techniques whereby classifiers (such as random decision forest or Bayesian classifiers) are trained in a supervised manner to detect cells of interest.

The main weakness of these techniques is handling data that differ even slightly from the training set. The myriad variance that you see in imaging data not only include intensity differences but morphological ones as well. Shallow learning classifiers for instance often fail to accurately detect tumor cells (typically large and cuboidal nuclei) and invasive immune cells (small and elongated nuclei) in the same image. It is in this variable space that DL approaches excel. Classifiers can be "trained by example" to capture the range of variance represented by the data and, furthermore, the addition of customizable data augmentation allows for even more robust classifiers to be developed.

It is clear from the direction of the field that the future of cell detection will certainly involve DL. While some of the original architectures for biological image segmentation are robust and still in use [24], other approaches that take advantage of the tremendous possibilities of DL will stand to further improve the options available. One interesting recent approach is the use of DL and star convex polygons to detect objects [26]. This technique has been shown to perform well across a range of modalities, however, it can break down for more complex structures or cells when the objects are not easily represented as star-convex objects.

4.1.4.3. Phenotypic classification

Pure data problems such as the phenotypic clustering discussed in Section 2.3 are more amenable to machine learning approaches than the semantic spatial segmentation discussed above. Indeed, given the typically Gaussian nature of both background and signal components, a Gaussian Mixture Model generally works well. In some cases, a multimodal approach can be used, however, it is rare to truly classify cells in this way (for example, as differentially intermediate positive and strong positive) as opposed to simply positive or negative for a probe.

With that said, the very concept of using the intensity alone to phenotype cells can be limiting (Figure 11). One example where more of the information contained within the image can be leveraged is in looking at

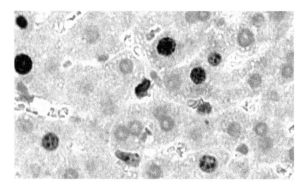

Figure 11:   An example of variable texture can be seen in this IHC image. Note that the DAB (dark brown) chromogen presents either in puncta or as a more homogenous stain. This can be relevant when phenotyping cells and would not necessarily be captured when looking at the stain intensity only.

not intensity *per se* but in the texture of the intensity. By using metrics such as variance, entropy or other Fourier-based texture features, biologically relevant metrics can be captured.

An example of this approach is used in Cell Painting [6], a machine learning technique that utilizes hundreds of features for phenotypic profiling. The logical extension of this idea is to also include morphological properties such as form factor and size, as well as spatial properties. The latter, however, are complicated by the fact that a single tissue section is inherently spatially anisotropic; that is, you have a different amount of information in the lateral (XY) plane than you do in the axial (Z) plane relative to the plane of sectioning. Regardless of the feature space, the main disadvantage of these techniques is that despite the huge dataset for profiling, without any analytics (during or after image analysis), the profile created may be meaningless. The weighting of only the most important features is, again, one of the pillars of a Convolutional Neural Network approach to clustering and phenotyping, with the main downside being that one needs *a priori* knowledge of your cells of interest on which to train a network.

4.1.4.4. Accessible data visualization and exploration

While the complex technologies involved in the image and data processing described above are critical to the development of robust tools for multiplex phenotyping, it is important to also consider the data exploration tools that allow visualization and interaction with the results. This will be critically important to drive these types of analyses into the realm of precision medicine.

Improvements to data visualization not only include creating relevant phenographs to provide information required by an end user, but also to develop tools that allow an intuitive interactive approach. For example, being able to identify clusters of cells in complex phenographs such as t-SNE plots (see Figure 7) and subset these for further analysis or reporting. We predict that this bi-directional approach will be critical in complex companion diagnostic biomarkers discovery where high-plex data are required at early stages to validate complex relationships. As the field matures, it is clear that new approaches to data visualization with a special emphasis on interactivity will play an important role.

4.1.5. *Summary*

This section has tried to convey both the limits of techniques presently seen in the field and also some of the cutting-edge approaches that are being taken. The overwhelming theme here is one of information density. Techniques, be they reagent, hardware, or software based, aim to increase the amount of information accessible from an image or dataset. In many cases, the sources of information are yet to be discovered, however, top-down DL approaches, where the networks are trained to provide a specific result with less emphasis on how they get there, serve an excellent proof of concept that the information exists.

4.2. *Clinical practice*

There is great hope to bringing histological multiplexing into clinical practice as more information can be provided to clinicians using less tissue from the biopsy. Chromogenic multiplexing has penetrated the clinical practice. Virtual Double Staining [19] is an approved clinical technique that uses the co-registration of adjacent serial section stained for panCK + p63 double IHC and ER, PR, or Ki67 IHC, respectively. The combination of these biomarkers in a virtual slide allows the pathologists to visualize the tumor areas (panCK), pre-invasive lesions (p63), and biomarker positive/negative cells (ER, PR, Ki67) simultaneously. This AI-assisted technique increases the speed and reproducibility with which pathologists can read breast biopsies.

There are examples of IF multiplexing used in the clinic mostly through FISH, but, in general, IF multiplex has not penetrated clinical practice except at a few sites that have created their own Lab Developed Test (LDT). However, more and more studies suggest that understanding not only the number of cells but also their cellular phylogeny is vital for understanding disease and treatment. The Stevenson Lab at UTMB recently published a paper using multispectral imaging to differentiate 19 unique macrophage populations in patients with autoimmune hepatitis, hepatitis C virus, and non-alcoholic steatohepatitis [25]. The paper argued that a deeper understanding of the macrophage populations through multiplex imaging platforms will allow for personalized treatment in patients depending on the phenotypes expressed in their livers.

IMC is the newest multiplexing technology and currently resides in the research and discovery realm. However, the clinical benefit is not too far away. Understanding the state of even more biomarkers on tissue biopsies in relation to the disease will help to refine disease categories and personalize treatments. It also provides an opportunity to monitor biomarkers outside of the disease to establish baseline or possibly probe for other pathologies. The challenge will be interpreting the massive amount of data these instruments will produce and simplifying for diagnostic purposes. Many discoveries made by IMC will be distilled into more simplistic assays for the clinic.

References

[1] Akturk, G., Sweeney, R., Remark, R., Merad, M., and Gnjatic, S. (2020). Multiplexed immunohistochemical consecutive staining on single slide (MICSSS): Multiplexed chromogenic IHC assay for high-dimensional tissue analysis. In *Biomarkers for Immunotherapy of Cancer. Methods in Molecular Biology*, Thurin, M., Cesano, A., and Marincola F. (eds.), Vol. 2055, New York, NY: Humana, doi: 10.1007/978-1-4939-9773-2_23.

[2] Amir, el-A. D., Davis, K. L., Tadmor, M. D., Simonds, E. F., Levine, J. H., Bendall, S. C., Shenfeld, D. K., Krishnaswamy, S., Nolan, G. P., and Pe'er, D. (2013). viSNE enables visualization of high dimensional single-cell data and reveals phenotypic heterogeneity of leukemia. *Nat. Biotechnol.* 31(6): 545–552, doi: 10.1038/nbt.2594.

[3] Baharlou, H., Canete, N. P., Cunningham, A. L., Harman, A. N., and Patrick, E. (2019). Mass cytometry imaging for the study of human diseases-applications and data analysis strategies. *Front. Immunol.* 10: 2657, doi: 10.3389/fimmu.2019.02657.

[4] Becht, E., McInnes, L., Healy, J., Dutertre, C. A., Kwok, I. W. H., Ng, L. G, Ginhoux, F., and Newell, E. W. (2018). Dimensionality reduction for visualizing single-cell data using UMAP. *Nat. Biotechnol*, doi: 10.1038/nbt.4314.

[5] Bolognesi, M. M., Manzoni, M., Scalia, C. R. *et al.* (2017). Multiplex staining by sequential immunostaining and antibody removal on routine tissue sections. *J. Histochem. Cytochem.* 65(8): 431–444.

[6] Bray, M. A., Singh, S., Han, H., Davis, C. T., Borgeson, B., Hartland, C., Kost-Alimova, M., Gustafsdottir, S. M., Gibson, C. C., and Carpenter, A. E. (2016). Cell painting, a high-content image-based assay for morphological profiling using multiplexed fluorescent dyes. *Nat. Protoc.* 11(9): 1757–1774, doi: 10.1038/nprot.2016.105. Epub 2016 Aug 25.

[7] Camp, R., Chung, G., and Rimm, D. (2002). Automated subcellular localization and quantification of protein expression in tissue microarrays. *Nat. Med.* 8, 1323–1328, https://doi.org/10.1038/nm791.

[8] Carvajal-Hausdorf, D. E., Patsenker, J., Stanton, K. P., Villarroel-Espindola, F., Esch, A., Montgomery, R. R., Psyrri, A., Kalogeras, K. T., Kotoula, V., Foutzilas, G., Schalper, K. A., Kluger, Y., and Rimm, D. L. (2019). Multiplexed (18-Plex) measurement of signaling targets and cytotoxic T cells in trastuzumab-treated patients using imaging mass cytometry. *Clin. Cancer Res.* 25(10): 3054–3062, doi: 10.1158/1078-0432.CCR-18-2599.

[9] Chang, Q., Ornatsky, O. I., Siddiqui, I., Loboda, A., Baranov, V. I., and Hedley, D. W. (2017). Imaging mass cytometry. *Cytometry A* 91(2): 160–169, doi: 10.1002/cyto.a.23053.

[10] Francis, J. E., Mason, D., and Lévy, R. (2017). *Beilstein J. Nanotechnol.* 8: 1238–1249, doi:10.3762/bjnano.8.125.

[11] Freiberg, B. and Baird, R. (2021). Biomarker colocalisation analysis of a virtual 12-plex using discovery chromogenic dyes and tissuealign™ co-registration software, Visiopharm poster. Available at https://cdn2. hubspot.net/hubfs/3461014/Knowledge%20Base%202020/Downloadables/ Poster-20180807-Biomarker-colocalisation-analysis.pdf.

[12] Gall, J. G. and Pardue, M. L. (1969). Formation and detection of RNA-DNA hybrid molecules in cytological preparations. *Proc. Natl. Acad. Sci. U.S.A.* 63(2): 378–383, doi: 10.1073/pnas.63.2.378.

[13] Galland, C., Ghosh, Y., Steinbrück, A. *et al.* (2011). Two types of luminescence blinking revealed by spectroelectrochemistry of single quantum dots. *Nature* 479: 203–207, https://doi.org/10.1038/nature10569.

[14] Grimm, J., English, B., Chen, J. *et al.* (2015). A general method to improve fluorophores for live-cell and single-molecule microscopy. *Nat. Methods* 12: 244–250, https://doi.org/10.1038/nmeth.3256.

[15] Johnson, J., Kasprzak, A., Chen, T., Nguyen, J., and Bui, M. (2017). Research methodologies and applications in WSI analysis: A perspective from an analytic microscopy core laboratory. *J. Pathol. Inform.* 8: 46.

[16] Keren, L., Bosse, M., Marquez, D., Angoshtari, R., Jain, S., Varma, S., Yang, S. R., Kurian, A., Van Valen, D., West, R., Bendall, S. C., and Angelo, M. (2018). A structured tumor-immune microenvironment in triple negative breast cancer revealed by multiplexed ion beam imaging. *Cell* 174(6): 1373–1387.e19, doi: 10.1016/j.cell.2018.08.039.

[17] Lehtoviita, A., Rossi, L. H., Kekäläinen, E., Sairanen, H., and Arstila, T. P. (2009). The CD4(+)CD8(+) and CD4(+) subsets of FOXP3(+) thymocytes differ in their response to growth factor deprivation or stimulation. *Scand. J. Immunol.* 70(4): 377–383, doi: 10.1111/j.1365-3083.2009.02307.x.

[18] Lin, J. R., Izar, B., Wang, S., Yapp, C., Mei, S., Shah, P. M., Santagata, S., and Sorger, P. K. (2018). Highly multiplexed immunofluorescence imaging of human tissues and tumors using t-CyCIF and conventional optical microscopes. *Elife* 7: e31657, doi: 10.7554/eLife.31657.

[19] Lykkegaard Andersen, N., Brügmann, A., Lelkaitis, G., Nielsen, S., Friis Lippert, M., and Vyberg, M. (2018). Virtual double staining: A digital approach to immunohistochemical quantification of estrogen receptor protein in breast carcinoma specimens. *Appl. Immunohistochem. Mol. Morphol.* 26(9): 620–626, doi: 10.1097/PAI.0000000000000502.

[20] Maaten, L. V. D. and Hinton, G. (2008). Visualizing data using t-SNE. *J. Mach. Learning Res.* 9: 2579–2605.

[21] Method of the year 2020: Spatially resolved transcriptomics. (2021). *Nat. Methods* 18(1): 1, doi: 10.1038/s41592-020-01042-x.

[22] Remeniuk, B., Coltharp, C., Roman, K., Wang, C., Neumeister, V., and Hoyt, C. (2020). Developing and validating a high-throughput 9-color immunofluorescence assay for translational studies in immuno-oncology research [abstract]. In *Proceedings of the Annual Meeting of the American Association for Cancer Research; 2020 Apr 27–28 and Jun 22-24*, Philadelphia (PA): AACR; *Cancer Res.* 2020; 80(16 Suppl): Abstract nr 2831, doi: 10.1158/1538-7445.AM2020-2831.

[23] Røge, R., Riber-Hansen, R., Nielsen, S., and Vyberg, M. (2016). Proliferation assessment in breast carcinomas using digital image analysis based on virtual Ki67/cytokeratin double staining. *Breast Cancer Res. Treat.* 158(1): 11–19, doi: 10.1007/s10549-016-3852-6. Epub 2016 Jun 9.

[24] Ronneberger, O., Fischer, P., and Brox, T. (2015). U-Net: Convolutional networks for biomedical image segmentation, https://arxiv.org/abs/1505. 04597 [cs.CV].

[25] Saldarriaga, O. A., Freiberg, B., Krishnan, S., Rao, A., Burks, J., Booth, A. L., Dye, B., Utay, N., Ferguson, M., Akil, A., Yi, M., Beretta, L., and Stevenson, H. L. (2020). Multispectral imaging enables characterization of intrahepatic macrophages in patients with chronic liver disease. *Hepatol. Commun.* 4(5): 708–723, doi: 10.1002/hep4.1494.

[26] Schmidt, U., Weigert, M., Broaddus, C., and Myers, G. (2018). Cell detection with star-convex polygons. In *Medical Image Computing and Computer Assisted Intervention — MICCAI 2018. MICCAI 2018. Lecture Notes in Computer Science*, Frangi, A., Schnabel, J., Davatzikos, C., Alberola-López, C., and Fichtinger, G. (eds.), Vol. 11071, Springer, Cham, https://doi.org/10.1007/978-3-030-00934-2_30.

[27] Ståhl, P. L., Salmén, F., Vickovic, S., Lundmark, A., Navarro, J. F., Magnusson, J., Giacomello, S., Asp, M., Westholm, J. O., Huss, M., Mollbrink, A., Linnarsson, S., Codeluppi, S., Borg, Å., Pontén, F., Costea, P. I., Sahlén, P., Mulder, J., Bergmann, O., Lundeberg, J., and Frisén, J. (2016). Visualization and analysis of gene expression in tissue sections by

spatial transcriptomics. *Science* 353(6294): 78–82, doi: 10.1126/science. aaf2403.

[28] Wolff, A. C., Hammond, M., Allison, K. H., Harvey, B. E., Mangu, P. B., Bartlett, J., Bilous, M., Ellis, I. O., Fitzgibbons, P., Hanna, W., Jenkins, R. B., Press, M. F., Spears, P. A., Vance, G. H., Viale, G., McShane, L. M., and Dowsett, M. (2018). Human epidermal growth factor receptor 2 testing in breast cancer: American Society of Clinical Oncology/College of American Pathologists Clinical Practice guideline focused update. *J. Clin. Oncol.: Official J. Am. Soc. Clin. Oncol.* 36(20): 2105–2122, https://doi. org/10.1200/JCO.2018.77.8738.

Chapter 7

An Introduction to Deep Learning in Pathology

Jeremias Krause[*,§]**, Heike I. Grabsch**[†,‡,¶]**, and
Jakob Nikolas Kather**[*,‡,‖]

*Department of Medicine III, University Hospital RWTH Aachen,
Aachen, Germany*

†*Department of Pathology, GROW School for Oncology and
Developmental Biology, Maastricht University Medical Center*[+]*,
Maastricht, The Netherlands*

‡*Division of Pathology and Data Analytics, Leeds Institute of Medical
Research at St. James's, University of Leeds, Leeds, UK*

§*jeremias.krause@rwth-aachen.de*
¶*h.grabsch@maastrichtuniversity.nl*
‖*jkather@ukaachen.de*

1. Introduction

Digitized histopathology images contain an immense number of visual features. Attempts to automatically extract information from these images have traditionally relied on manually designed methods in specialized applications. In the last years, however, computer-based image analysis approaches have been markedly improved by machine learning methods, in particular deep learning. Deep learning relies on artificial neural networks which are able to automatically detect patterns in digitized images. For example, in the context of solid tumors, deep learning can not only detect cell types or other tissue structures, but can also directly predict survival or other high-level concepts from raw image data. While a number of promising studies have been published, many pathologists and researchers still struggle to make sense of this tool. This chapter aims to provide a guide for them on how to get started with deep learning, by presenting a basic workflow and important concepts for the application of deep learning in digital histopathology.

2. Basic Applications of Machine Learning in Digital Pathology

In the context of digital pathology, machine learning can be a helpful tool to automate repetitive and simple tasks. Prime examples are counting tumor-infiltrating lymphocytes [1], scoring Ki-67-stained tissue slides [2], or detecting tumor tissue in tissue slides of axillary lymph nodes in breast cancer patients [3]. These problems are well-defined and can be easily solved by a trained pathologist but can be very time-consuming. While for some of these tasks, automated computer-based approaches were available even before the deep learning era [4], nowadays a range of publications have shown that computers can solve these simple tasks with a very good performance which is potentially even usable in clinical routine [5]. We have previously summarized some of these tasks in a review of the literature [6]. All of these digital image processing approaches rely on pathology slides being digitized in clinical routine. While this is not yet the case in most countries, digital workflows can be expected to be the norm in one or two decades. Digitalization in medicine impacts clinical workflows and needs immense resources to be implemented. Yet, nowadays almost all radiology departments rely on completely digital workflows. This precedent demonstrates that a medical specialty can successfully undergo such a transformation from conventional to digital workflows.

3. End-to-End Deep Learning Biomarkers in Digital Pathology

Using machine learning to automate simple, tedious tasks in daily pathology practice could save time and resources for pathologists.

However, these applications do not encompass the full extent of possible applications of machine learning in pathology. Exceeding these simple tasks, machine learning methods have been used for "end-to-end" image analysis tasks. This means the detection of a clinically applicable metric (e.g., a prognostic or predictive score) directly from raw digitized histopathology slides, without human intervention during the learning process. For example, deep learning networks can be directly trained on image data to predict patient survival, thereby constituting an end-to-end biomarker [7]. A range of studies have shown that deep learning can predict clinical endpoints such as survival or disease recurrence in a number of tumor types [8–10]. Another example for end-to-end deep learning workflows are methods to detect molecular alterations directly from histopathology images [11–13]. In many types of cancer, genetic tests are run in addition to histological examination of slides to determine the mutational status of one or multiple genes in tumor tissue (such as EGFR or ALK in non-small cell lung cancer) [14]. Other molecular tests are run on tumor tissue to detect overexpression of specific proteins (such as hormone receptors in breast cancer) or more complex genetic changes in tumor tissue, such as microsatellite instability in colorectal cancer [15, 16]. It has been repeatedly shown that deep learning is capable of detecting mutations directly from hematoxylin-and-eosin-stained (H&E) slides [14, 15, 17, 18], suggesting that genetic subtypes of cancer differ in their morphological presentation.

By being able to predict mutations directly from H&E slides, deep learning has the potential to speed up the diagnostic process. In summary, machine learning methods could speed up current workflows in pathology, but could also enable completely new biomarkers [6]. Consequently, it is helpful for pathologists and researchers to understand the technological basis enabling this progress. In the following sections, we will give a brief technical introduction about the technology behind basic and end-to-end deep learning applications in image processing, specifically in the context of digital pathology.

4. Neural Network Technology Basics

Machine learning is an approach in computer science to develop programs which are able to find patterns in complex datasets without explicit guidance. A particularly relevant example is image data. While images are easily understandable to a human observer, storing an image in a computer results in a large set of numbers with no inherent structure. In the previous decades, automatically analyzing images with computer programs was regarded among the hardest tasks in machine learning and required meticulous manual processing of the source data. Today, even very difficult image analysis problems can be easily solved by machine learning. One prominent example of machine learning methods which has enabled this progress is the deep learning artificial neural network. These algorithms can be trained using data — e.g., histological images — to recognize specific patterns and classify new samples (e.g., differentiating tumor from normal tissue). To achieve this, data with known labels (e.g., images of tumor and non-tumor tissue) are used to train the network and to validate the results (Figure 1).

The architecture of neural networks mimics the function of biological networks: they process complex (visual) data and produce high-level outputs. To be exact, an artificial neural network is a set of so-called nodes connected to each other. In analogy to the brain, each node is called a neuron. A subset of nodes organized to a column is called a layer. There are different types of layers fulfilling different tasks, e.g., an input and an output layer. Applying neural networks to pathology implies an input layer capable of reading raw image data.

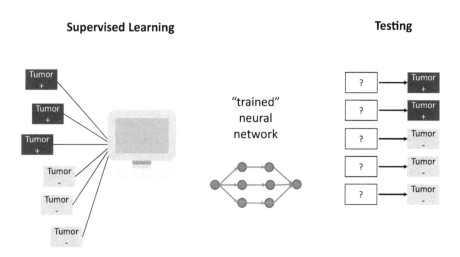

Figure 1:   Simplified workflow of supervised deep learning.

Notes: In supervised learning, the real label of the data is available to the neural network during all steps of the training process. After the neural network is trained, it is able to distinguish between the labels.

A computer interprets images as a matrix or array that stores individual pixel values. Images are processed into tiles usually containing a height and width of 512 pixels, respectively, as well as three color channels (red, green, and blue; RGB). This results in a $512 \times 512 \times 3$ matrix for each tile of the image. These can now be fed into the network sequentially. Each time the network processes a new image tile during training, it learns from it. To be exact, the connections between neurons are changed by a very small degree during each step of the training process. In machine learning jargon, the "parameters" of a network are "updated" during training.

5. Image Data Flow in Deep Neural Networks

When the data flows through the neural network, it is first processed by the input layer. Apart from taking the data into the network, the input layer applies different data normalization steps to the matrix, which is explained in more depth elsewhere [19]. After this step, the input layer transmits the data to the first hidden layer. Hidden layers describe all the layers between input and output. Neural networks with more than two hidden layers are called "deep" neural networks, referring to their increased complexity of model architecture. The training of the network occurs in the hidden layers, as most of them slightly modify the data received according to parameters (or "weights"), some of which are preset (hyperparameters) and some of which change with every step during the training process.

The general data flow inside a neural network follows the forward propagation method: The output of one neuron is the input for the next neuron. Each neuron processes its input data with an "activation function". This function is nonlinear, which means that it is not just a multiplication of the input with another number. This enables the layers to identify complex patterns in the input data. The activation function processes the input data together with a learned value, its "weight". Applying the weights usually refers to multiplying the weight with the input from one prior neuron and to add up all the products coming from connected neurons.

After passing the hidden layers, the processed data arrives at the output layer. Here, a final decision or label is given to the data based on weighted predictions from each layer. While at first glance, this cascade can seem complicated, in reality it is just a concatenation of millions of small and simple computation steps. During training, the true label of each item in the training data is revealed to the neural network and the network calculates the magnitude of error for the individual prediction. In technical jargon, this error is called the "loss". The process of calculating this error is mathematically embedded in a "loss function". Having mathematically defined this loss function, a technique called back-propagation will change all parameters in the neural network to reduce this error after each training step. This is how an artificial neural network "learns" from input data. After each learning step, the network will reach a slightly more accurate prediction. Calculating a loss and then changing parameters to decrease that loss is a common problem in mathematics and engineering. A range of simple mathematical tools are available to solve such "optimization" problems in a computationally efficient way. Often, these optimization methods work in an iterative step-by-step manner. Instead of spotting the perfect solution right away, a step-by-step solution needs to be employed to solve the optimization problem with a neural network. In the context of neural networks, a commonly applied optimization algorithm is the gradient descent method. To make neural networks understandable, their architecture can be compared to biological networks. However, the details of training artificial neural networks are very different from training a biological neural network. While it is often said that artificial neural networks function like real biological neurons, artificial neural networks utilize some mechanisms which most likely play no significant role in biological neurons (e.g., back-propagation) [17]. Intuitively, while an artificial neural network has to make thousands or millions of small computational steps to detect simple patterns, a biological neural network can remember a new object after just seeing it once.

6. The Art of Applying Machine Learning

Sometimes, the process of training neural networks to a new type of data is referred to as an "art" instead of a purely scientific or engineering endeavor. The main reason for this is that during the training process, a number of design choices have to be made by the researcher. Collectively, these design choices and computer settings are referred to as "hyperparameters". Hyperparameters can be described as the frame of conditions that control how the training of the neural network is conducted and are therefore central to the deep learning process. For example, a very important hyperparameter is the "learning rate" which influences the rate of change of the weights during the training process. To determine optimal hyperparameters for the problem at hand is a key question in deep learning and often relies on experimenting and intuition. A practical approach is to look up recommended values inside published papers and also to try out your own ideas. Trying out your own ideas relies on some basic understanding of computer programming, which is easily attainable even for non-specialists. In recent years, a number of online platforms and programming packages such as "fastai" [20] have made coding accessible for anyone who is motivated to invest a few weeks in learning the basics of computer programming. Even without programming tools, some software tools allow us to apply artificial neural networks to histopathology images for non-programmers. For example, QuPath [21], an open-source software tool for digital histopathology, supports tissue classification with artificial neural networks and other state-of-the-art machine learning algorithms.

7. Data Pre-processing in a Digital Pathology Pipeline

Step-wise workflows for image processing are called "pipelines". In the case of deep learning analysis of whole-slide images, a consensus pipeline has emerged from dozens of independent publications, comprising multiple steps [22]. For pathologists and researchers who have the motivation and time to familiarize themselves with the basics of computer programming, we will give practical advice on how to set up a deep learning experiment.

The journey of the data starts at the patient with the extraction of tissue that is suspected to be a potential tumor. After processing the tissue in pathology (embedding, cutting, staining), the tissue slides can be scanned and analyzed on electronic devices. Succeeding the scanning, the data are stored on a hard drive in one of multiple conventional formats (e.g., SVS, TIFF, NDPI) using different types of data compression (JPEG, JPEG2000, or others). The next step in a deep learning project is the pre-processing of the data. We will consider a case where tissue slides are cut into small tiles, which are then presented to a neural network in order to predict an outcome for the respective patient: Histological whole-slide images are "gigapixel" images — they usually contain more than one billion (10^9) pixels. Current deep neural networks are optimized for images in the range of ten thousand pixels (10^4). Therefore, histological images need to be cut into smaller parts before analysis. For the annotation and tiling process, the open-source software QuPath [21] can be used, but a range of other tools are also available. After gathering the tiles, the images are usually color normalized, for example with Macenko's method [23]. Most of the time, a filter is applied to exclude tiles without useful information or tiles containing a lot of distraction (noise). The data groups should be balanced, either by leaving out data from the majority class (undersampling) or by reusing data from the minority class (oversampling).

To increase the model's generalization ability, data augmentation is used to increase variety. Common methods include randomly flipping or rotating the images or zooming into random image regions. Methods like stretching, on the other hand, could tamper with biological features and are therefore not recommended.

After pre-processing the image data, it is then sorted into three groups: training, testing, and validation (e.g., in a ratio of 70-15-15). Most applications in digital pathology are supervised approaches, which means that a gold standard (or "ground truth") information is available during training. In supervised learning, the data are prelabeled and the network learns to distinguish data according to that label (Figure 1). Before being

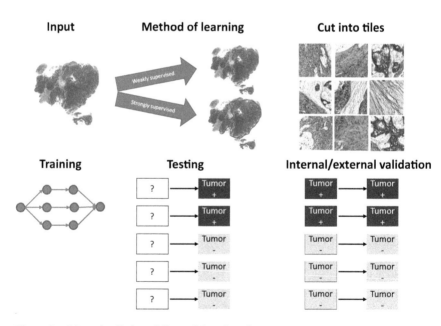

Figure 2:   More detailed workflow of deep learning.

Notes: After pre-processing the data, choosing the method of deep learning, and adjusting the parameters, training, testing, and validation steps are conducted.

used as input data for the artificial neural network, the data usually need to undergo a series of refinements (Figure 2).

8. Hardware and Computing Requirements

To get started, deep learning does not require a vast amount of data for training, testing, and validating the neural network. Typically, a few dozens of digitized slides contain enough information for proof-of-concept workflows. However, a requirement to iteratively develop deep learning methods is to have access to sufficient computing power (Figure 3). In most cases, this does not require huge computing clusters, but a modern desktop workstation computer equipped with a graphics card (graphics processing unit, GPU). Alternatively, researchers can experiment in cloud computing platforms such as Google Colab [24] which provide immediate access to sufficient computing power. Neural networks are usually trained on GPUs: due to their number of graphic processors these devices can enable a parallel processing training environment [25]. Multiple GPUs can be used together to speed up training. The correlation between the number of GPUs and the speed of the training is positive though not exponential. For most digital pathology problems, a single GPU is sufficient to achieve state-of-the-art results. After choosing the training environment, data need to be acquired and organized. Often the data contain information on multiple levels: the patient-, slide-, and tile-level (Figure 4).

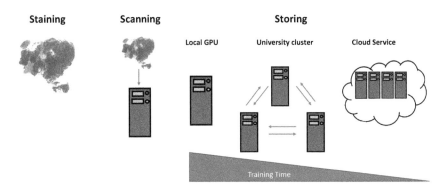

Figure 3:   Technical requirements for deep learning.

Notes: As described in the text, to speed up the training process, enough hardware capacity is essential.

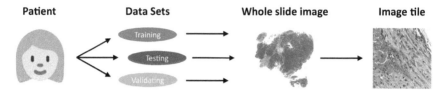

Figure 4:   Important steps of data processing.

Notes: Patient-, slide-, and tile-levels are displayed. The available data are split into training, testing, and validation sets. Note that one patient may contribute multiple slides and therefore even more tile images.

A single patient can contribute more than one slide and one slide typically contributes thousands of small image tiles. This is important to keep in mind to prevent the network from just memorizing which tiles belong to which patient, instead of learning biologically relevant patterns. This aspect makes working with digital pathology data different than in other applications and requires caution throughout the development process.

9. Choosing and Training a Neural Network Model

Following the setup process, a neural network itself is needed. The most demanding way would be to create a neural network from scratch. For almost all practical applications, re-using a pre-existing network is preferable and yields better results in less time. Usually, deep learning networks have been pre-trained on the *image classification challenge* [26–28]. These networks can be re-trained on new histology data, which can be much more efficient than training from scratch. This process is called transfer learning and is a successful strategy in many deep learning applications. When using transfer learning, not all layers need to be trained: often, only the deepest layers which are close to the output layers are trainable; or, in technical jargon, "hot". Consequently, an interesting hyperparameter for this form of deep learning is the number of hot layers. Only these layers are updated through the back-propagation process while layers earlier in the process responsible for low-key features like lines or edges are frozen. Applying this method speeds up the training process significantly [29]. However, using a pre-trained network also has some disadvantages: The user is running a risk of using a model which is trained to be too complex and contains too many parameters for a relatively easy classification task. By properly modifying the pre-trained network, these issues usually can be avoided. The final step to make a prediction is to add up the individual tile predictions and decide to either classify the whole slide image as, e.g., positive or negative. An easy but not always sufficient way to do this is to calculate the arithmetic mean and use that as an indicator for the whole case. Another important part of the analysis is to test the model's generalizability; this is best tested by presenting a never seen dataset to the trained model (Figure 2). A model trained on a too narrow dataset or trained for too many epochs on the same dataset can suffer from something called "overfitting", most probably only memorizing the dataset but not able to transfer its knowledge to a new dataset. If the model is not able to learn the features, it might be too complex or too shallow and it might be better to train a smaller or larger network.

There are several pitfalls in deep learning which should be avoided in order to achieve a good performance (Figure 5). For example, it is important to have an equal distribution between the different labels, otherwise the neural network tends to use the more frequent label. This problem is called class imbalance and can be very relevant for digital pathology tasks. Also, the original data should not be stored in very high compression to avoid compression artifacts diluting relevant image information. Finally, a practical problem relevant in digital pathology workflows is the so-called batch effect. This phenomenon describes the effect that slides undergoing the preparation process at the same local place share some hidden features with each other.

In summary, to get started with deep learning, a basic understanding of a few governing principles is necessary, but most of the intuition for

Batch effect   **Signal to noise ratio**   **Compression and zooming artifacts**

Figure 5:   Common pitfalls.

Notes: The batch effect leads to false prediction due to non-label-related similarities in tiles preprocessed at the same institution. A high signal-to-noise ratio distracts the neural network from the relevant information stored in the tissue. Compression and zooming artifacts distort the results.

running experiments has to be acquired during experimenting with tasks of growing complexity. Fortunately, today, several graphical interfaces and easy-to-master programming tools are available to enable this journey not only for computer scientists, but for anybody willing to invest the time to get hands-on-experience in deep learning.

10. Explainable Deep Learning

What can humans learn from deep learning? To this date, it is still mostly unclear which image features are really relevant for the neural network [30], the network might literally see histological slides from a completely different angle than humans do. For the most part, deep learning models a "black box" so it is difficult to understand the network's predictions right away. However, some methods have been developed to visualize individual features the network learned or to visualize which regions of a specific tile lead to the network's decision. Furthermore, using deep learning to predict molecular features from HE pictures [15] could be used to identify cases where a pronounced intratumor heterogeneity was detected. By using feature visualization techniques [31] one might discover features in tumor slides so far not recognizable by the human eye.

11. Outlook

To this day, a lot of preclinical work has been done to show that machine learning has valuable implications for the medical field [6]. It can speed up daily workflows (e.g., determination of molecular subtypes of tumors) and will most likely be used as a decision support system inside the clinic. However, most published approaches are still on an academic level and not ready to be directly applied in clinical workflows. Developing promising deep learning workflows from the academic proof-of-concept level to clinically usable products with regulatory approval is not a trivial task. However, ultimately, deep learning could have the potential to improve daily practice of pathology and possibly improve patient outcomes. Key actors for these future developments are multidisciplinary and include pathologists, researchers, clinicians, engineers as well as other academic and industrial partners. Training all relevant actors in basic deep learning histopathology will serve as a basis for the necessary dialog and further developments.

References

[1] Galon, J., Costes, A., Sanchez-Cabo, F., Kirilovsky, A., Mlecnik, B., Lagorce-Pagès, C. *et al.* (2006). Type, density, and location of immune cells within human colorectal tumors predict clinical outcome. *Science* 313: 1960–1964.

[2] Saha, M., Chakraborty, C., Arun, I., Ahmed, R., and Chatterjee, S. (2017). An advanced deep learning approach for Ki-67 stained Hotspot detection and proliferation rate scoring for prognostic evaluation of breast cancer. *Sci. Rep.* 7: 3213.

[3] Ehteshami Bejnordi, B., Veta, M., Johannes van Diest, P., van Ginneken, B., Karssemeijer, N., Litjens, G. *et al.* (2017). Diagnostic assessment of deep learning algorithms for detection of lymph node metastases in women with breast cancer. *JAMA* 318: 2199–2210.

[4] Kather, J. N., Weis, C.-A., Bianconi, F., Melchers, S. M., Schad, L. R., Gaiser, T. *et al.* (2016). Multi-class texture analysis in colorectal cancer histology. *Sci. Rep.* 6: 27988.

[5] Kleppe, A., Skrede, O.-J., De Raedt, S., Liestøl, K., Kerr, D. J., and Danielsen, H. E. (March 2021). Designing deep learning studies in cancer diagnostics. *Nat. Rev. Cancer* 21(3): 199–211, doi: 10.1038/s41568-020-00327-9. Epub 2021 Jan 29.

[6] Echle, A., Rindtorff, N. T., Brinker, T. J., Luedde, T., Pearson, A. T., and Kather, J. N. (February 2021). Deep learning in cancer pathology: A new generation of clinical biomarkers. *Br. J. Cancer* 124(4): 686–696, doi: 10.1038/s41416-020-01122-x. Epub 2020 Nov 18.

[7] Calderaro, J. and Kather, J. N. (2020). Artificial intelligence-based pathology for gastrointestinal and hepatobiliary cancers. Gut. p. gutjnl–2020, doi:10.1136/gutjnl-2020-322880.

[8] Skrede, O.-J., De Raedt, S., Kleppe, A., Hveem, T. S., Liestøl, K., Maddison, J. *et al.* (2020). Deep learning for prediction of colorectal cancer outcome: A discovery and validation study. *Lancet* 395: 350–360.

[9] Kather, J. N., Krisam, J., Charoentong, P., Luedde, T., Herpel, E., Weis, C.-A. *et al.* (2019). Predicting survival from colorectal cancer histology slides using deep learning: A retrospective multicenter study. *PLoS Med.* 16: e1002730.

[10] Kather, J. N. and Calderaro, J. (October 2020). Development of AI-based pathology biomarkers in gastrointestinal and liver cancer. *Nat. Rev. Gastroenterol Hepatol.* 17(10): 591–592, doi: 10.1038/s41575-020-0343-3.

[11] Fu, Y., Jung, A. W., Torne, R. V., Gonzalez, S., Vöhringer, H., Shmatko, A. *et al.* (2020). Pan-cancer computational histopathology reveals mutations, tumor composition and prognosis. *Nat. Cancer* 1: 800–810, https://doi.org/10.1038/s43018-020-0085-8.

[12] Schmauch, B., Romagnoni, A., Pronier, E. *et al.* (2020). A deep learning model to predict RNA-Seq expression of tumours from whole slide images. *Nat. Commun.* 11: 3877, https://doi.org/10.1038/s41467-020-17678-4.

[13] Kather, J. N., Heij, L. R., Grabsch, H. I., Loeffler, C., Echle, A., Muti, H. S. *et al.* (August 2020). Pan-cancer image-based detection of clinically actionable genetic alterations. *Nat. Cancer* 1(8): 789–799, doi: 10.1038/s43018-020-0087-6. Epub 2020 Jul 27.

[14] Coudray, N., Ocampo, P. S., Sakellaropoulos, T., Narula, N., Snuderl, M., Fenyö, D. *et al.* (2018). Classification and mutation prediction from non–small cell lung cancer histopathology images using deep learning. *Nat. Med.* 24: 1559–1567.

[15] Kather, J. N., Pearson, A. T., Halama, N., Jäger, D., Krause, J., Loosen, S. H. *et al.* (2019). Deep learning can predict microsatellite instability directly from histology in gastrointestinal cancer. *Nat. Med.* 25: 1054–1056.

[16] Echle, A., Grabsch, H. I., Quirke, P., van den Brandt, P. A., West, N. P., Hutchins, G. G. A. *et al.* (2020). Clinical-grade detection of microsatellite instability in colorectal tumors by deep learning. *Gastroenterology* 159(4): 1406–1416.e11, doi: 10.1053/j.gastro.2020.06.021. Epub 2020 Jun 17.

[17] Kermany, D. S., Goldbaum, M., Cai, W., Valentim, C. C. S., Liang, H., Baxter, S. L. *et al.* (2018). Identifying medical diagnoses and treatable diseases by image-based deep learning. *Cell* 172: 1122–1131.e9.

[18] Saillard, C., Schmauch, B., Laifa, O., Moarii, M., Toldo, S., Zaslavskiy, M. *et al.* (December 2020). Predicting survival after hepatocellular carcinoma resection using deep-learning on histological slides. *Hepatology* 72(6): 2000–2013, doi: 10.1002/hep.31207.

[19] Goodfellow, I., Bengio, Y., Courville, A., and Bengio, Y. (2016). *Deep Learning*, Cambridge: MIT Press.

[20] Howard, J. and Gugger, S. (2020). Fastai: A layered API for deep learning. *Information* 11: 108.

[21] Bankhead, P., Loughrey, M. B., Fernández, J. A., Dombrowski, Y., McArt, D. G., Dunne, P. D. *et al.* (2017). QuPath: Open source software for digital pathology image analysis. *Sci. Rep.* 7: 16878.

[22] Muti, H. S., Loeffler, C., Echle, A., Heij, L. R., Buelow, R. D., Krause, J. *et al.* (2020). The aachen protocol for deep learning histopathology: A hands-on guide for data preprocessing. *Zenodo*, doi:10.5281/ZENODO.3694994.

[23] Macenko, M., Niethammer, M., Marron, J. S., Borland, D., Woosley, J. T., Guan, X. *et al.* (2009). A method for normalizing histology slides for quantitative analysis. In *2009 IEEE International Symposium on Biomedical Imaging: From Nano to Macro*, Boston, MA, USA: IEEE, pp. 1107–1110.

[24] Kanani, P. (August 2019). Deep learning to detect skin cancer using Google Colab. *International Journal of Engineering and Advanced Technology (IJEAT)* 8(6): 2249–8958.

[25] Sakhare, S. (2015). *Parallalize Neural Network Training with GPU Computing*. Sacramento: California State University.

[26] Deng, J., Dong, W., Socher, R., Li, L., Li, K., and Fei-Fei, L. (2009). ImageNet: A large-scale hierarchical image database. In *2009 IEEE Conference on Computer Vision and Pattern Recognition*, Miami, FL, USA: IEEE, pp. 248–255.

[27] Russakovsky, O., Deng, J., Su, H., Krause, J., Satheesh, S., Ma, S. *et al.* (2015). ImageNet large scale visual recognition challenge. *Int. J. Comput. Vis.* 115: 211–252.

[28] Krizhevsky, A., Sutskever, I., and Hinton, G. E. (2012). Imagenet classification with deep convolutional neural networks. *Adv. Neural Inf. Process. Syst.* 25: 1097–1105.

[29] Ilin, R., Watson, T., and Kozma, R. (2017). Abstraction hierarchy in deep learning neural networks. In *2017 International Joint Conference on Neural Networks (IJCNN)*, pp. 768–774.

[30] Rudd, E. M., Günther, M., Dhamija, A. R., Kateb, F. A., and Boult, T. E. (2018). What's hiding in my deep features? *Deep Learning Biometrics* 153–174, doi:10.1201/b22524-7.

[31] Olah, C., Mordvintsev, A., and Schubert, L. (2017). Feature Visualization. *Distill* 2, doi:10.23915/distill.00007.

Chapter 8

AI-Driven Precision Pathology: Challenges and Innovations in Tissue Biomarker Analysis for Diagnosis

**Dirk Vossen[*,‡], Jeppe Thagaard[*,§], Fabian Schneider[*,¶],
Rasmus Norré Sørensen[*,‖], Johan Doré[*,**],
Esther Abels[*,††], Amanda Lowe[*,‡‡], Mogens Vyberg[†,§§],
and Michael Grunkin[*,¶¶]**

[*]*Visiopharm A/S. Agern Allé 24, 2970 Hørsholm, Denmark*

[†]*Aalborg University Copenhagen, Copenhagen, Denmark*

[‡]*dvo@visiopharm.com*
[§]*jth@visiopharm.com*
[¶]*fsc@visiopharm.com*
[‖]*rns@visiopharm.com*
[**]*jdh@visiopharm.com*
[††]*eab@visiopharm.com*
[‡‡]*alo@visiopharm.com*
[§§]*mov@dcm.aau.dk*
[¶¶]*mgr@visiopharm.com*

1. Basic Concepts from Image Analysis, AI, and Digital Pathology

Digital pathology is the disruptive technology that refers to the use of computer technology to generate, view, analyze, and manage digitized anatomic pathology tissue samples. Here, we focus on introducing some of the basic concepts regarding analyzing pathology images using artificial intelligence (AI), often referred to as image analysis, as well as cover key terms around digitalization of the pathology laboratory.

To make image analysis applicable in the research and clinical lab, it is important to consider how it is integrated seamlessly into digital workflows that allow scientists and clinicians to work efficiently. Thus, digital precision pathology must incorporate considerations around digital workflows, storage, standardization, and regulatory pathways. These topics are also briefly reviewed in the following sections.

1.1. *Digitalization and whole-slide images — the foundation of AI*

Since Bacus Research Laboratories filed their first patent on virtual microscopy back in 1998 [4], more than 30 companies now produce whole-slide scanners. Conceptually, a whole-slide scanner captures images of the glass slide at high magnification and stitches these together in a pyramidal digitized format called a whole-slide image (WSI), see Figure 1.

WSI scanners enable pathologists to view sections on digital monitors [24], but also open up for the quantitative analysis of micro-structural content, including biomarker response, using computer-aided methods. Beyond brightfield WSI scanners, there are now also fast fluorescent systems that enable other ways of visualizing structures and functions of tissues and cells, as can be seen in Figure 2. Lately, the use of AI/deep learning has revolutionized the ability to analyze such image datasets.

1.2. *Basics of image analysis and AI*

The fundamental aspect of image analysis in digital pathology is that it must provide quantitative and reproducible interpretation to support the

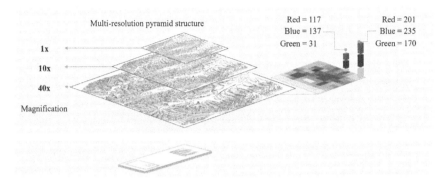

Figure 1:   A typical WSI used for diagnostic reading is scanned at 40x (0.25 μm/pixel) or 20x (0.50 μm/pixel) magnification, generating a giga-pixel image (~200.000 × 100.000 pixels) which is stored in a multi-resolution pyramid structure. The resolution and file format ensure that is comparable to an optical microscope in terms of image access, e.g., zooming, panning, etc.; a brightfield image is a three 8-bit color channel representing the red, green, and blue colors that are used in RGB brightfield images. For comparison, modern smartphones capture images containing 4032 × 3024 pixels RGB images, i.e., in terms of pixels, a single WSI includes the same data as more than 1500 iPhone images.

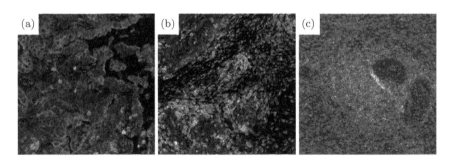

Figure 2:   Imaging modalities: Beyond the brightfield WSI, research laboratories use multiplexed immunofluorescent WSIs with 3–12 individual color channels (a: Akoya Biosciences, MA, USA; b: Ultivue, MA, USA) up to high-plex imaging mass cytometry WSIs with more than 40 color channels (c: Fluidigm, CA, USA). Each color channel represents an individual biomarker contrary to RGB colors of brightfield imaging.

users with accurate and reliable data, which is often critical to making important diagnostic-, research-, or even business decisions.

In the following, we describe the key differences between traditional approaches to image analysis and AI/deep learning-based approaches, but first, we will briefly review the basic concepts that enable us to go from a digital image to quantitative output results as these steps apply to both approaches. These steps include [9]:

(1) *Pre-processing* includes aspects such as image normalization, image smoothing to remove noise, etc., where the image information is not changed significantly.
(2) *Feature engineering* involves applying different filters to enhance certain signals in the image, e.g., a blob-filter enhances the signal for round objects in the image, or image deconvolution, where the DAB signal in the image is isolated from the haematoxylin. These steps usually involve knowledge and training as it is the experience of the user that drives the quality of the generated features by knowing which steps enhance what signals in the image.
(3) *Classification* is the process of assigning a certain class to each entity,[1] e.g., pixels in the image, so we can start forming the relevant objects in the image. Most importantly, we generate the rules that go from the input (pre-processed image or feature image) to the segmentation output, enabling us to delineate the outline of certain histology objects in the image. These rules are usually learned by the computer, hence the term machine learning, but in simple cases, the user can also manually set these.
(4) *Post-processing* is a powerful step in any image analysis algorithm, where the output of the classification can be further processed, and object-level heuristic and objective rules can easily be incorporated into the analysis pipeline, e.g., removing small nuclei below a certain area or other histology relevant rules that apply to the problem at hand.
(5) *Output calculation* is usually the last step of any image analysis algorithm as until now we have not quantified the objects from the classification and post-processing steps. Most importantly, these are usually human interpretable formulas, e.g., connectivity of the membranes, counting of CD8+ nuclei, measuring the area of the tumor, and also the combination of these that outputs the density of CD8+ nuclei inside the tumor as a number per mm^2.

These concepts are written here as high-level as possible as any image analysis algorithm can be described by either one or more combinations

---

[1]Classification can also be aimed at other entities such as already segmented objects or an entire field-of-view.

of these. The configurability of these depends only on the image analysis platform and the creativity of the user.

As described above, an image analysis algorithm involves the translation of many histology or image processing rules into a language that the computer can understand and execute to obtain the desired output. In a traditional approach, many of these are engineered by the user, and with machine learning, the classification rules can be learned by the computer using a small amount of annotated training data. Whereas this approach is generally applicable to, e.g., industrial inspection of well-defined objects, it has severe limitations when it comes to the complexity and variability of biological structures. Variability exists at every conceptual level of the analysis problem, and usually, that cannot be reduced to a finite set of exact and enumerated rules that fully covers the observed biological variability. This is one of the reasons that classical machine learning approaches in computational pathology have generally suffered from a lack of robustness and/or complexity of the underlying algorithm (APP) developed for the application at hand. These challenges have knock-on effects on everything from user-experience, over data quality, and to a multitude of regulatory considerations and concerns.

Deep learning is a relatively new field of machine learning, and it is the underlying method that drives the new wave of AI in digital pathology. Relative to the classical framework discussed above, deep learning is an approach where the computer learns many of the concepts only using annotated training data associated with the image. Most importantly, it combines the feature engineering and classification concepts, replacing most of the difficult and tedious work of translating rules into the computer while still being complemented by the rest of the concepts. This means that by providing annotated training data of what the ideal output should be, the computer learns to recognize patterns and translates that into the rules that produce, e.g., the segmentation output. This approach represents an evolution from the original artificial neural networks. It has resulted in nothing less than a breakthrough in computational pathology in two critical dimensions: (1) The scope of analysis applications appears to be unlimited where problems that were close to impossible to tackle with classical methods (e.g., H&E tumor detection, mitotic figure recognition, etc.) are easily and robustly addressed with deep learning, and (2) the general robustness and reliability of the algorithms.

Thus, the shift toward deep learning-based algorithms opens new possibilities as it surpasses the performance of traditional approaches on difficult tasks. But it also introduces new challenges, including data access, training, hardware, and entirely new regulatory considerations and paradigms. Some of these topics will be discussed in the following sections.

When it comes to choosing the right approach to the analysis of WSIs, it is probably worth remembering the words of Oscar Wilde, *the truth is rarely pure and never simple* (*The Importance of Being Earnest*, 1895). Most real-life applications of image analysis in pathology exist across several conceptual levels and scales. There is usually always the need to identify both regions (e.g., tumor vs. stroma), identify objects (e.g., cells), and characterize and manage heterogeneity (to find e.g., hot-spots). Application knowledge and interpretation of the segmented image is often based on input from the pathologist, where spectral, relational, and contextual information is converted into rules that are applied before the quantification of output parameters. All of these steps will usually need to incorporate image information across several levels of magnification (i.e., levels in scale-space). In practical terms, that means that we will typically need both AI-based approaches alongside more classical rule-based approaches to arrive at the most efficient and robust solutions. Therefore, we strongly believe that the future belongs to those who can efficiently integrate and unify all of these approaches, thus offering the optimal choice of approach for each conceptual aspect of the analysis application.

1.3. *Standardization and interoperability for analysis purposes*

Like radiology, the WSI market started with a multitude of proprietary file formats, blocking the free exchange and manipulation of WSIs. Likewise, this has also created a demand for standardization through DICOM. The DICOM committee formed *Working Group 26*[2] to facilitate the adoption of digital pathology imaging into hospitals and laboratories by standardizing the image formats and communication. Today, DICOM with *Supplement 145: Whole-Slide Microscopic Image* is the preferred and recommended standard for storing and communicating brightfield WSIs in clinical laboratories. Picture Archiving and Communication Systems (PACS) and Vendor Neutral Archives (VNAs) are evolving into becoming

---

[2] https://www.dicomstandard.org/activity/wgs/wg-26 (accessed on March 30, 2021).

de-facto standards for workflows in diagnostic and research labs. And just as in radiology, these systems support DICOM standards. In reality, however, the support still has many gaps that challenge both data access and general interoperability between systems. This challenge is also well known from radiology, and to some extent caused by narrow and short-term commercial considerations.

Both for image analysis in general and the high-plexed complex image formats, a standard still has a long way to come. It is however encouraging that more and more whole-slide vendors seem to adopt the BigTIFF-based open standard for WSIs formed by the *Open Microscopy Environment (OME)*[3] effort. The missing consensus on a standard is also reflected in the PACS, VNAs, and Image Managements Systems (IMSs) available to the research laboratories, as every single vendor has its proprietary approach to storing and communicating WSIs.

With the advent of AI and image analysis for digital pathology, the lack of standardization in storing and communicating quantitative results and annotations is further exacerbated. This is an often-forgotten aspect of standardization in digital pathology, and only recently the DICOM committee has started working on such standardization. To encompass image analysis and AI, we recommend saving image analysis results as polygons and not separate WSI-based formats. Polygons can represent the image overlays such as nuclei, membranes, cytoplasm, tumor regions, etc. This greatly simplifies calculation on object relation and the support of analysis at various magnifications. This is well-aligned with the latest recommendations by the DICOM committee, however, full standardization across both storing of image and analysis results is yet to be seen but will be critical for interoperability for both clinical and research environments.

2. Tissue Biomarkers in the Era of Precision Medicine

2.1. *Introduction*

In the dawning era of personalized medicine, companion diagnostic (CDx) biomarkers have become critical gatekeepers to the development

---

[3] https://www.openmicroscopy.org (accessed on March 30, 2021).

and prescription of new personalized drugs. They have become a requirement for patient stratification in clinical trials, regulatory approval, market access, and treatment selection once the drug is approved. The search for CDx biomarkers spans a range of different technology platforms, where gene sequencing and molecular diagnostics technologies are new promising technologies. However, there is also an increasing demand for tissue diagnostic methods, typically based on immunohistochemistry (IHC), where tissue biomarkers have shown great promise with their ability to offer multidimensional and spatially resolved information, thus providing the context necessary for understanding the tumor microenvironment (TME).

Biopharmaceutical companies will typically consider CDx biomarkers along at least four dimensions.

- *Platform*: The technology platform should aid systematic approaches to the discovery of new biomarkers.
- *Predictive*: The biomarker has to be precise and offer the predictive power required for correct patient stratification in clinical development and treatment selection.
- *Deployment*: There must be an economic and scalable path into routine diagnostic use.
- *Coverage*: The (companion) diagnostic test must be easily available to a meaningful fraction of patients.

Tissue biomarkers have the potential to offer powerful solutions along these four dimensions. Existing examples of tissue biomarkers used in current routine diagnostics include the assessment of human epidermal growth factor 2 (HER2) status and the expression level of estrogen receptor/progesterone receptor (ER/PR) for invasive breast cancers using IHC assays. While these and other tissue biomarkers are part of standard protocols for treatment selection, there are several challenges related to variation in the preparation and interpretation of the tissue samples that must be mitigated. We will discuss the types and magnitudes of these challenges, as well as potential solutions. The solutions considered here are based on tools that use AI and Image Analysis (IA) of digital images from tissue samples created using Digital Pathology (DP) tools. Some of these solutions are well tested and proven, while others are in the proof-of-concept stage. With this overview, we want to demonstrate and recommend a path forward for mitigating current limitations in the use of IHC assays.

2.2. *The quality challenge with tissue biomarkers*

Before a tissue sample from a patient can be evaluated by a pathologist to come to a diagnostic, prognostic, or predictive result, the material must undergo a series of manual and semi-automated steps to create a formalin-fixed paraffin-embedded (FFPE) tissue section, which is then stained with haematoxylin-eosin (HE), a special stain and/or IHC. These steps include obtaining the material (biopsy or resection specimen), fixation, grossing, processing, embedding, sectioning, staining, and cover slipping of the tissue sample. In each of the steps, there are sources of variation and potential sources of error. Also, the diagnostic assessment of the tissue sample by a pathologist is associated with considerable inter- and intra-observer variability. The magnitude depends on the specific endpoint, the complexity of the case and/or biomarker in question. Figure 3 shows the major sources of variation and error, along with the approximate magnitude and impact as reported in the literature.

Modern pathology laboratories have processes and controls in place to mitigate several of these errors, with a stated goal of standardizing the diagnostic process. These include quality systems, the use of controls, continuous training, and enrolment in external quality assessment (EQA) programs. In the following sections, we will highlight several quality gaps

Process Steps	Biopsy and fixation	Slide Preparation	Staining	Digitization	Diagnostic evaluation
Sources of variation & error	▪ Up to 20% of biopsies unfit for Dx ▪ Delayed, under- or over-fixation	▪ Folds, tears ▪ Cells/tissues lost or mixed-up	▪ 20%-30% of labs have insufficient stain quality for Dx	▪ Missing tissue ▪ Out of focus ▪ Dark nuclei	▪ Variability in interpretation of CDx biomarkers (e.g. PD-L1, Ki67, HER2)
Impact	Longer turn-around time, re-work in the lab, complexity of Dx process Delayed / incorrect diagnostic, Sub-optimal / incorrect patient treatment, Patient outcomes				

Figure 3: Process steps in the assessment of tissue samples, the associated sources of variation and error for each step and examples of the impact of these issues on the patient, the healthcare system, and economics.

that we have an opportunity to close with AI-based methods, in the pursuit of the level of standardization required for tissue diagnostics to unfold its true potential in precision medicine.

2.3. *Toward standardized assessment of tissue biomarkers*

The different types of assessment errors in histology and IHC manifest themselves in different ways for different biomarkers and their specific diagnostic applications. We have selected three concrete examples to highlight more general underlying problems that typically challenge the reproducibility, sensitivity, and specificity of biomarker assessment by pathologists using a microscope.

In each case, we will point to potential solutions using AI. Of these, some have already demonstrated their robustness, and have been adopted in routine diagnostic workflows. Other potential solutions are still being developed but indicate directions that should be pursued for tissue diagnostics to truly enter the era of precision medicine.

2.3.1. *Invasive tumor detection and management*
of tumor heterogeneity

2.3.1.1. Background

The use of Ki67 in routine diagnostics for breast cancer is a good practical example for highlighting how limitations in the human cognitive system can impact the ability to correctly classify cells as, e.g., invasive tumor cells, stromal cells, pre-invasive cells, or (proliferating) tumor-infiltrating lymphocytes. Heterogeneity with respect to biomarker response is often observed, and for Ki67, this can be quite significant. Scientific studies have demonstrated the importance of reading Ki67 in the hottest hot-spot to achieve the maximum prognostic power of this biomarker (e.g., [34]), and reading of this biomarker in hot-spots has been adopted in clinical guidelines. Unfortunately, it has also been demonstrated that less than 50% of pathologists are able to correctly identify the hottest hot spot [3], introducing significant variability and challenges to the diagnostic utility of such manual reads. It remains unclear whether a hot-spot is a spot with high concentration or a high proportion of Ki67 positive cells. Using

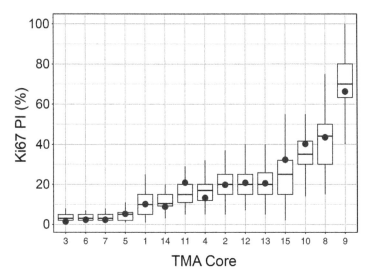

Figure 4: Inter-observer variability between almost 200 human readers compared to automated APP (gray dot) of Ki67 PI across different TMA cores. Adapted from Ref. [32].

Ki67 as an example, we can quantify the magnitude of the impact these error sources have on inter- and intra-reader variability [32] (see Figure 4) and ultimately on diagnostic misclassification. The results and observations here can likely be extrapolated to other biomarkers, where visual classification of cells and tumor heterogeneity are challenged. Indeed, we expect this to be the case for new CDx biomarkers, where the visual complexity is even higher.

2.3.1.2. Magnitude of the problem

The high inter- and intra-observer variability of Ki67 scoring is due to intratumoral heterogeneity and the current practice of using manual eye-balling [29, 28, 37]. This has a direct impact on the diagnostic accuracy for separation of Luminal B from A subtypes of breast cancer leading to higher misclassification of patients [35], where as much as 30% diagnostic misclassification was observed when using the PAM50 gene-expression assay as a reference.

2.3.1.3. Proposed solutions

This type of variability can be overcome by using digital image analysis as shown by multiple publications to be an objective and reproducible alternative to manual scoring. Røge *et al.* showed that an automated Ki67 approach could significantly lower the inter-observer variability found in individual samples scored by 199 pathologists [32].

Limiting the biomarker analysis only to the invasive tumor cells is a key element of an automatic approach, i.e., an algorithm should have the ability to exclude normal gland epithelium and pre-invasive lesions. Hida *et al.* proposed virtual triple staining for digital scoring of Ki67 using consecutive sections of Ki67, pan cytokeratin (panCK), and p63 to automatically exclude pre-invasive lesions, e.g., ductal and lobular carcinoma *in situ* (DCIS, LCIS) [12]. An alternative that has also been used in clinical practice is virtual double staining, where one section is physically double-stained with panCK and p63 (or p40) (see Figure 5). These approaches are simple, robust, and verifiable, and allow for non-pathologists to run the technical aspects of the analysis and may even enable full automation of the technical aspects of the analysis, prior to review by a pathologist. But this comes at the added costs of extra staining and increased complexity in the lab.

More recently, AI deep learning-based approaches for global Ki67 scoring on whole-slide images (WSI) have shown great promise to streamline the fully automated analysis to only using the primary biomarker-stained slide [14]. Further validation and regulatory approval are required for this to be used in a routine clinical context. But this approach has already demonstrated its usefulness in a range of research applications.

To tackle intra-tumor heterogeneity, Wessel Lindberg *et al.* proposed a novel approach to objectively calculate the Ki67 PI index in hot-spots [41]. This method builds on the virtual double staining method [31] so that both tumor and Ki67 heterogeneity could be automatically assessed in a quantitative and reproducible manner. This methodology has been validated in clinical practice on WSIs [18] and shown to outperform manual biomarker assessment for both Luminal B classification and PAM50 gene assay agreement [35]. Especially digital hot-spot analysis is superior to both systematic manual Ki67 counting and mitotic figure (pHH3) counting, adding significantly higher prognostic values [34], see Figures 6 and 7. This could be due to the fact that visual and digital hot-spot analyses are

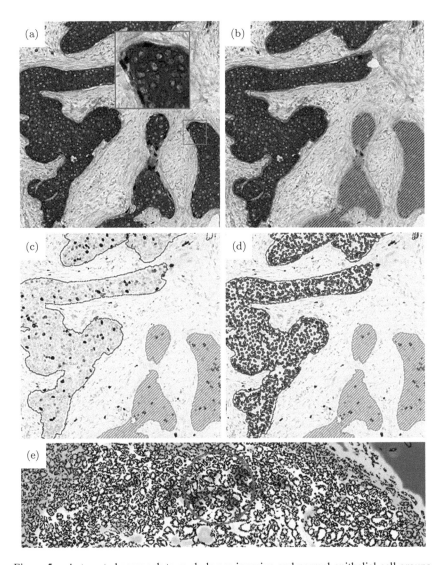

Figure 5: Automated approach to exclude pre-invasive and normal epithelial cell groups. One section is physically double-stained with panCK and p63 (a), where the APP (Visiopharm #20101, www.visiopharm.com) uses the p63+ to discriminate between invasive (blue outline) and pre-invasive epithelial regions (orange outline) (b). The serial section with Ki67 is aligned with the first section (c), transferring the results and enabling precise quantification of Ki67 positive nuclei only in the invasive tumor (d). The analysis is done on the entire slide; hence we can assess the local Ki67 PI index and perform objective hot-spot quantification according to the geographical region's clinical guidelines (e).

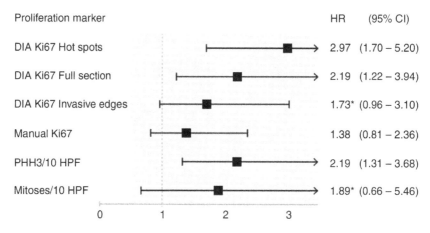

Figure 6: Cox regression hazard ratios for all-cause mortality within 10 years of breast cancer diagnosis. Adapted from Ref. [34].

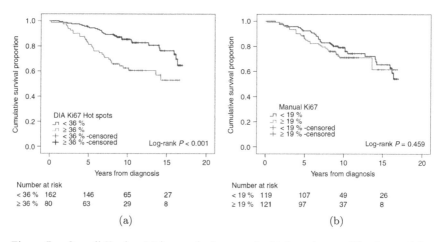

Figure 7: Overall Kaplan–Meier survival curves for high vs. low proliferating activity from digital image analysis (a) and manual assessment (b). Adapted from Ref. [34].

different: in the former, the pathologist's attention is dragged to fields (of different sizes) with a high number of Ki67 positive cells, which can either represent a high concentration of cells or a high Ki67 index. In a digital hot-spot analysis, we analyze local densities of negative and positive cells across the entire slide and calculate the Ki67 PI in a hot-spot with a defined number of cells. Thus, the true hot-spots can be identified and measured. Recently, Robertson *et al.* suggested global Ki67 scoring as an alternative to hot-spot analysis in terms of overall survival if performed by digital image analysis [30].

2.3.1.4. Summary

The accuracy, reproducibility, and clinical utility of even a simple routine biomarker such as Ki67 requires a lot of complex steps, as discussed above: (1) correct identification of invasive tumor components, exclusion of pre-invasive components and stromal cells, (2) identification of the hottest hot-spot, and (3) accurate counting and classification of cells within hot-spots. The same or very similar challenges are pre-requisites to standardization of biomarker assessment for (companion) diagnostic tissue biomarkers, which will only continue to increase in complexity compared to the routine diagnostic biomarkers used today. And, in turn, standardization will be essential for tissue diagnostics to enter the realm of precision medicine. With the increased volume of diagnostic tests, and the increased complexity of those tests, AI-based tools are required to achieve automation in diagnostic workflows to avoid a severe diagnostic bottleneck in the healthcare system. This will be a challenge given the lack of standardization and a scalable integration approach across image formats, image acquisition parameters, reagent platforms, and digital infrastructure components. Some of these challenges will be discussed further in the following sections.

2.3.2. *The transition from qualitative to data-driven quantitative pathology*

2.3.2.1. Background

Clinical guidelines and diagnostic cut-offs are essential to the practice of standardized precision medicine. Such guidelines are typically established based on a multitude of comprehensive studies, and evolve through scientific discourse between scientists and experts in their respective fields. But much of the data that such guidelines and cut-offs are based on are either qualitative or semi-quantitative, and as such may lack sensitivity, specificity, and reproducibility as discussed both in this chapter and in countless scientific publications. AI and image analysis are tools that allow us to explore quantitative morphometric endpoints for microstructural properties in general, and biomarker expression in particular. That, in turn, allows us to discover, develop, and validate quantitative endpoints that are optimized with respect to predictive, prognostic, or diagnostic purposes. The example used in the following is for HER2 and its application for treatment selection for breast cancer patients. We will demonstrate how AI and image analysis have enabled robust quantification of membrane morphometry, which may even be more predictive of treatment effect than the current Gold Standard of HER2-FISH. This example is also used to demonstrate the potential of AI-driven quantitative pathology to not only enhance the power of tissue biomarkers but also to standardize their practical application.

2.3.2.2. The problem

HER2 overexpression/amplification occurs in 15–20% of primary breast adenocarcinomas [33, 43, 45], and is a predictive biomarker for anti-HER2-therapy, e.g., trastuzumab. It is usually assessed in daily routine through a two-tiered approach; first with IHC graded as negative (0 [no staining or incomplete faint membrane staining of ≤10% of cells] or 1+ [incomplete faint membrane staining in >10% of tumor cells]), equivocal (2+ [weak to moderate complete membrane staining in >10% of tumor cells]), or positive (3+ [strong complete membrane staining]) (see Figure 8). Equivocal cases are secondly assessed by ISH, based on the

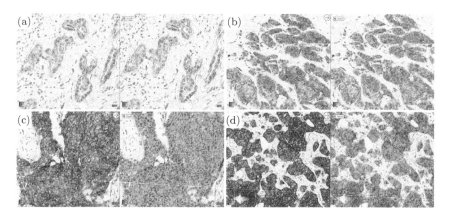

Figure 8:   Representative HER2 cases. (a) Manual = 1+, APP (Connectivity) = 1+ (0.10), (b) Manual = 2+, APP (Connectivity) = 1+ (0.19), ISH = negative, (c) Manual = 3+, APP (Connectivity) = 3+ (0.68), ISH = positive, (d) Manual = 3+, APP (Connectivity) = 3+ (0.94).

ratio of HER2 gene copy number per number of centromeric regions of chromosome 17 (CEP17): A ratio of ≥2 is considered positive. Also a gene copy number per tumor cell ≥6 is considered positive. Manual HER2 IHC assessment is hampered by large inter-observer variability with only 70–80% agreement when giving a binary (0, 1+ vs. 2+, 3+) score on the same set of cases with equal distribution of scores [42]. Likewise, the intra-observer variability is high with ~80% agreement when reviewing the same set of cases multiple times with a set washout period of two weeks [42].

The manual IHC assessment tends to lead to a high number of equivocal (2+) cases, hence adding cost and turnaround time by increasing the number of 2+ cases that require HER2 ISH testing.

Already in 2007, the American Society of Clinical Oncology (ASCO) and College of American Pathologists (CAP) acknowledged the need for and recommended the use of quantitative image analysis, especially for cases showing weak HER2 membrane staining (ASCO/CAP Guidelines 2007) [44]. With the updated ASCO/CAP guidelines in 2013 and 2018, the HER2 score should be seen as the relationship between the intensity and the completeness of circumferential membrane staining (ASCO/CAP Guidelines 2013, 2018).

The ASCO/CAP guidelines remain qualitative in the description of staining intensity and degrees of circumferential staining and do not guide as to the relative importance of intensity vs. membrane morphology. Also, they do not address how to handle membranous structures, where the associated nucleus is not present in the section. It is also not discussed how membranes that appear common to adjacent cells should be handled. Whereas pathologists may be able to holistically integrate this type of information, digital algorithms need exact answers.

In the wake of the ASCO/CAP guidelines in 2007, and the emergence of digital pathology, several commercial HER2 algorithms were developed and released, and some were even cleared by the FDA [7]. Most of the algorithms were optimized against manual reads, as opposed to, e.g., HER2-FISH or clinical outcomes. They were often developed for specific assay manufacturers. Due to heavy reliance on membrane intensity, there was typically a lack of robustness and translation between algorithms developed for different assays. Finally, these early algorithms did not incorporate membrane morphology. As a result, these first-generation algorithms exhibited relatively low sensitivity/specificity with respect to HER2-FISH. And the practical use remained low.

2.3.2.3. Proposed solutions

As a digital approach, Brügmann *et al.* (2012) [6] proposed *connectivity* as a novel approach to quantifying the intensity and membrane completeness using HER2-CONNECT™ for HER2 in invasive breast carcinoma [6]. They defined connectivity as a continuous measure for the positively stained membrane based on the size distribution of membrane components. In a multiple regression model against HER2-FISH, it was shown that membrane connectivity alone remained statistically significant. As membrane intensity did not contribute statistically independent information, it could be eliminated from the model. That in turn, allowed the same cut-offs to be used for assays across all the different vendors and labs.

The connectivity was shown to highly correlate with categorical HER2 scoring and has been also shown to generalize beyond breast carcinoma to other cancer types such as gastric and gastroesophageal junction adenocarcinomas [26, 19]. Furthermore, several studies have validated and proven the value of this approach on independent and external breast cancer cohorts [21, 17, 10] with a high correlation with the manual pathologist scoring and the HER2 ISH gene copy number [21, 10].

Several groups have shown the potential of using the HER2-CONNECT™ algorithm to reduce the number of 2+ cases [15, 10] by up to 75% without sacrificing sensitivity and specificity, potentially leading to lower cost and faster turnaround times in HER2 testing. Currently, HER2 ISH testing serves as the gold standard for HER2 status as a quantitative score of the HER2 gene. However, current therapies target the HER2 protein and not the gene. Here, Li *et al.* (2020) [22] recently used HER2-CONNECT™ to quantify the HER2 protein expression and showed that HER2 connectivity had the strongest association with the pathologic complete response (pCR) compared to the HER2 gene copy number and the HER2 chromosome 17 ratio for HER2-positive breast cancer patients receiving anti-HER2 neoadjuvant chemotherapy (Table 1, Figure 9) [22]. This shows the great potential of using new computational approaches to increase the power of HER2 as a companion diagnostic test.

2.3.2.4. Summary

Quantitative pathology offers the opportunity to develop robust, sensitive, and specific biomarker assessment which is superior in predictive,

Table 1: Multivariate analysis of factors associated with incomplete response to anti-HER2 neoadjuvant chemotherapy. Adapted from Ref. [22].

	pCR		Incomplete response			*p*
	#/median	%/range	#/median	%/range	OR (95% CI)	value
Case#	83	—	70	—	—	—
Age (years)	53	26–76	57	30–86	0.96 (0.91–1.00)	0.035
PR negativity	64	64%	36	36%	3.20 (1.24–8.59)	0.019
HER2 FISH copy number	19.9	3.6–34.36	12.46	4–35.5	1.07 (1.01–1.15)	0.040
HER2 DIA connectivity	0.88	0.5–0.97	0.75	0.12–0.98	136.08 (8.21–3965.94)	0.002

Notes: pCR pathologic complete response, *OR* odds ratio, *PR* progesterone receptor, *DIA* digital imaging analysis.

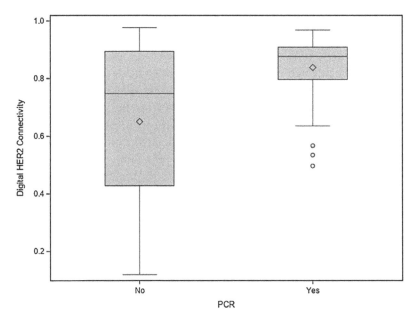

Figure 9: HER2 connectivity as a continuous biomarker scoring method for pathologic complete response (pCR). Adapted from Ref. [22].

prognostic, or diagnostic power when compared to manual assessment via qualitative or semi-quantitative guidelines. Also, it can be hard to translate clinical guidelines based on qualitative or semi-quantitative data directly into algorithms, which prolongs the time it takes to get widespread acceptance. We believe, and recommend, that quantitative pathology becomes more integrated into the development of clinical guidelines and standards, from the onset. This will require close collaboration between clinical practitioners, scientific communities, biopharma companies, assay manufacturers, regulators, and the developers of quantitative pathology software technology. Also, it will require access to large, curated datasets. As always, the challenge lies in the creation of an appropriate pre-competitive space where this type of exchange can happen. Such barriers will likely continue to limit both the availability and depth of AI-based diagnostic menus for the immediate future.

2.3.3. *Addressing the increased complexity and ambiguousness in clinical assessments of predictive biomarkers*

2.3.3.1. Background

In this section, we look at the potential future of predictive biomarkers in histopathology through the lens of the recent success of cancer immunotherapies. The development of novel drugs that target complex immune checkpoint pathways sets high requirements for the assessment of the tumor microenvironment (TME). Naturally, this can result in complex and ambiguous scoring guidelines for any associated companion diagnostic test. Here, we see AI-based tools as the promising but still untapped opportunity to obtain more knowledge with higher standardization. Probably the most successful immunotherapies are directed against PD-1 or its ligand PD-L1. PD-L1 IHC expression has become the first predictive biomarker for these immune checkpoint inhibitors approved by the USA's Food and Drug Administration (FDA). For PD-L1 IHC, many of the same observations from previous sections (Ki67, HER2) apply, however, the visual assessment in clinical routine involves more unmet needs in terms of heterogeneity and scoring methods. We use PD-L1 as an example due to its potential and relevance, but we also see a clear tendency that future biomarker candidates require more complex assessments of the TME with analysis of sub-cellular localization in various cellular phenotypes by utilizing multiplexed biomarker approaches.

2.3.3.2. The problem

Currently, there exist multiple drug-specific PD-L1 IHC assays as either companion diagnostics or as complementary diagnostic tests (Table 2). These assays are approved with different manual scoring guidelines to

Table 2: Overview table of the current complexity for PD-1/PD-L1 targeting drugs, IVD partner and IHC assays, associated scoring regime, cut-off values, and regulatory status to choose from for NSCLC.

Manufacturer	Drug target	Drug	IHC assay clone	IVD Partner	Cells scored	Cut-off	Regulatory status
Merck MSD	PD-1	Pembrolizumab	22C3	Dako/Agilent	Tumor cells	1%, 50%	Companion, PMA
BMS	PD-1	Nivolumab	28-8	Dako/Agilent	Tumor cells	1%, 5%, 10%	Complementary, PMA
AZ	PD-L1	Durvalumab	SP263	Ventana/Roche	Tumor cells	25%	Complementary, PMA
Roche	PD-L1	Atezolizumab	SP142	Ventana/Roche	Tumor cells and tumor-infiltrating lymphocytes	50% for TC 5% for TIL	Complementary, PMA
Merck KgAa & Pfizer failed for NSCLC	PD-L1	Avelumab	73-10	Dako/Agilent	Tumor cells and tumor-infiltrating lymphocytes	<1%, >1%, 50%, 80% for TC; STUDY FAILED	Companion, PMA

Notes: The Blueprint (BP) PD-L1 IHC assay project found that manually scored PD-L1 TC percentages were comparable between the assays 22-C3, 28-8, and SP263, whereas SP142 exhibited fewer stained TCs overall [13]. However, the results showed that 37% of cases would be scored differently depending on the assay/scoring system used and thereby would lead to misclassifications if the wrong assay would be applied. PD-L1 assay scoring for 22C3 and SP263 in NSCLC for clinically relevant cut-offs are not interchangeable [25]. Pathologists consistently scored 22C3 higher than SP263 on all 198 NSCLC TMA cores. Surprisingly, at 50% cut-off, around half of the cases positive with SP263 would have been defined negative with 22C3 by both pathologists [25].

semi-quantitatively assess the PD-L1 expression, e.g., using the assay 22C3 (pharmDX, Agilent) for identifying Non-Small Cell Lung Cancer (NSCLC)[4] patients for treatment with Pembrolizumab uses the Tumor Proportion Score (TPS):

$$TPS(\%) = \frac{\#PD-L1\,staining\,tumor\,cells}{Total\,\#\,of\,viable\,tumor\,cells} \times 100$$

where TPS is the percentage of viable tumor cells showing partial or complete membrane staining (≥1+) relative to all viable tumor cells present in the sample. There are several key challenges to PD-L1 assessment:

(1) Scale and heterogeneity: Require the manual assessment of all tumor cells (TC) in heterogeneous IHC patterns.
(2) Inflammation vs. tumor: Immune cells (IC) also express PD-L1 but scoring requires discrimination.
(3) Different assays, different cut-offs: Variability in expression and scoring guidelines across approved IHC assays.

These factors make the manual visual assessment difficult for pathologists and lead to higher intra- and inter-observer variability. This also makes it difficult for pathologists to choose the right antibody assay for cancer patients to be eligible for treatment.

Overall, pathologists showed strong reliability in TC scoring and poor reliability (high inter-observer variability) in IC PD-L1 scoring [36].

2.3.3.3. Proposed solutions

The difficulty of PD-L1 scoring and resulting variability in TC and IC scoring led to the development and publication of several digital image analysis solutions (uPath Roche Diagnostics) [1, 27, 16, 20, 11]. These methods vary in scoring approaches from academic pixel-based TPS scoring [16] to commercial CE-IVD-approved cell-level TPS scoring (SP263) with manual pathologist region annotations (uPath, Roche Diagnostics).

---

[4]For simplicity, we use Non-Small Cell Lung Cancer (NSCLC) as an example indication and discuss other indications in what follows.

The various PD-L1 scoring regimens for additional assays and indications (NSCLC, Urothelial carcinoma, Melanoma, Triple negative breast cancer) will require indication-specific algorithm development and validation for regulatory compliance (see Figure 10 for NSCLC example).

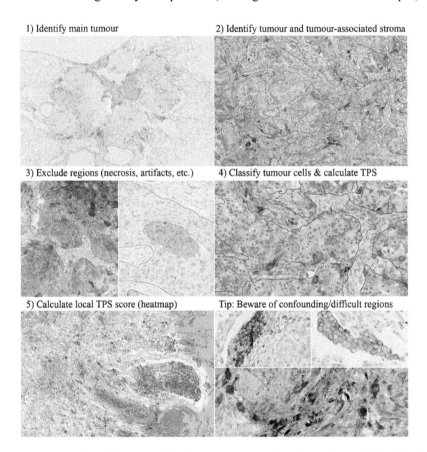

Figure 10:   Outline of a computational assessment procedure for scoring TPS in NSCLC. The main tumor (green) for assessment is automatically detected in (1), where invasive tumor (blue outline) is detected at higher power in (2). Necrosis and other non-quantifiable regions are also detected at higher power (3). Cells are then classified as tumor (dots) or non-tumor cells with the membrane PD-L1 expression determining if the cell is PD-L1 positive (red dot) or negative (blue dot) in (4). From this, we use the Visiopharm platform to calculate the global TPS score but can also visually show a local TPS score in (5) (red = high TPS, blue = low TPS). Depending on the assay, different areas might be difficult to discriminate from tumor cells which set high requirements for both training and validation data generation (Tip).

For example, the 22C3 IHC assay produced by Agilent may be used across a variety of indications beyond NSCLC that require a Combined Positive Score (CPS) to determine eligibility for treatment with Pembrolizumab:

$$CPS = \frac{PD-L1\,staining\,cells}{Total\,\#\,of\,viable\,tumor\,cells} \times 100$$

where the positive IC cells are combined with the positive tumor cells into a *combined* score. This scoring method considers the IC PD-L1 expression as increasingly established relevance similar to the IC/ICP scoring approach for the Ventana assays. However, as tumors differ in the relative importance of TC positive contribution and IC positive contribution, the CPS score captures both into a single fraction. It also makes it a notable candidate for digital automation as it requires precise evaluation along with a series of cell segregations, isolating DAB positive cells and then isolating tumor cells, so that quantitative measurements may be used in combination to determine the overall status of the sample objectively.

2.3.3.4. Summary

The success of immunotherapies targeting PD-(L)1 has generated four major clinical diagnostic IHC assays and scoring criteria. Selecting the right PD-L1 IHC assay and scoring has become a challenge for pathology laboratories. We need a better understanding of the interchangeability of the PD-L1 tests and their predictivity for response to anti-PD-(L)1 targeted therapy. This could be achieved with retrospective image analysis studies on large patient cohorts with known outcome data using the main PD-L1 IHC assays. This could validate current blueprint study results and provide guidance for pathology laboratories on assay and scoring method selection and pave the way for digital image analysis assistance for routine pathology. Translational biomedical studies have shown that understanding the patients' TME in immunotherapies is key for treatment response and have created several multiplex immunofluorescence (mIF) assays combining biomarkers such as CD3, CD8, CD20, CD68 with PD-1 and PD-L1 to analyze cellular phenotypes within their spatial distribution and neighborhood relations within the TME. The biggest challenge for pathologists to adopt this novel method is to review and QC these new types of images for translational medicine as to today's diagnostic brightfield IHC standard. So far, no standardized workflows for mIF QC and processing exist and their development will help to accelerate the adoption of this new method to translational and diagnostic pathology.

2.4. *Toward standardization in stain quality*

Staining quality and the reliability of tissue biomarker assessment has a significant impact on the interpretive accuracy of IHC-based diagnostic tests as discussed above. While the human costs of inaccurate diagnostic interpretation can be high, because of false negatives and false positives, so are the financial costs. In a recent study on HER2 staining conducted by the EQA organization Nordic Immunohistochemical Quality Control (NordiQC) and Roche, it was demonstrated that for every 1$ saved by using cheaper reagents in the pathology laboratory (i.e., based on in-house validated protocols instead of FDA-approved assays), the healthcare system is ultimately burdened with $6 in additional costs [38].

The magnitude of this problem is considerable and must be successfully addressed before tissue diagnostic biomarkers can achieve their full potential and enter the realm of precision medicine. From 2003–2015, NordiQC conducted 37 quality assessment runs in the General Module and 18 runs in the Breast Cancer Module. Over this period, a total of 89 epitopes have been part of up to 16 assessments and in total more than 30,000 IHC slides have been assessed. The overall results, as shown in Figure 11, show that about 30% of the stains in the general module and

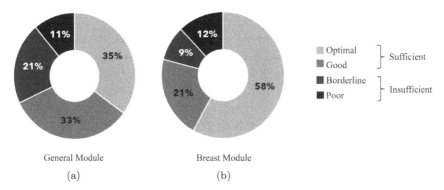

Figure 11: Pass and fail rates from NordiQC assessments between 2003 and 2015 for (a) General Module and (b) Breast Cancer Module. More than 30,000 slides from more than 500 laboratories have been assessed by an expert panel for the diagnostic quality.

20% in the Breast Cancer Module showed insufficient staining as assessed by NordiQC [39]. A later publication confirmed these numbers after the assessment of an additional 10,000 IHC slides [40], while it also concluded that the overall failure rate has not significantly changed over the previous 15 years. This is partially due to higher demands in the assessment (due to better antibodies, reagents, and stainers), but it remains remarkable that 20–30% of labs are not able to generate staining results that are sufficient for diagnostic use as assessed by a panel of expert pathologists in NordiQC. Similar results have been seen for other CDx markers, like for example PD-L1, where the average fail rate has been about 20% (Website NordiQC). Of all insufficient results, about 90% are due to weak or false negative staining, while 10% are due to a false positive or an overall too strong staining. Similar results across markers have been reported by other EQA organizations like UKNEQAS, CIQC, and CAP (website UKNEQAS, CIQC, CAP).

EQA organizations are addressing this important problem by offering regular proficiency testing schemes. Some schemes are also offering protocol review and recommendations in case of a failed test. While EQAs give laboratories valuable feedback on their performance regularly, there are typically at least 3 to 6 months between assessments due to bandwidth limitations at EQAs. For many biomarkers, quality runs are offered at a lower frequency, and for new biomarkers entering the market for new indications, test materials and schemes typically have to be developed after introduction. For the individual labs, there is no way to know how they perform in between EQA assessment runs. However, the variability in stain quality between quality runs can be considerable, as shown for PD-L1 in Figure 12.

Figure 12 shows an example of a laboratory (A) that can provide adequate staining in most runs. Laboratory (B) has a consistently lower score and several runs where they are below the threshold of sufficient staining, which is set at a score of 2.5. The third laboratory (C) has a good performance except for two runs where it scores below the threshold. These typical examples, from private communication with UKNEQAS, illustrate the variation found in the EQA assessments, where in any given run 20–30% of the labs' scores are below the threshold of sufficient quality for diagnostic interpretation. The variability between EQA runs represents a real problem in terms of standardization and a concomitant lack of robustness in tissue diagnostic testing. It goes

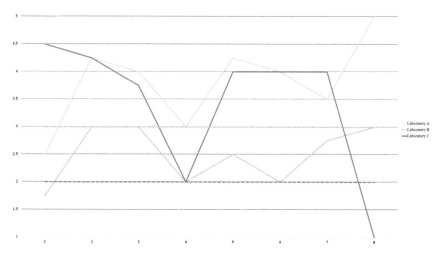

Figure 12:   Anonymized stain quality score for three pathology labs as assessed by expert pathologists for 8 consecutive PD-L1 EQA runs. The vertical score ranges from 5 (excellent), 4 (acceptable), 3 (borderline acceptable), and 2 and 1 (unacceptable). The thresholds for unacceptable and borderline acceptable are shown as a red and yellow dashed lines, respectively. Data come from private communication with UKNEQAS.

without saying that this is a problem both in routine diagnostics and in e.g., clinical trials.

There are several efforts to address this issue. One avenue is via using tissue controls. These controls are important for determining if a sample is over- or under-stained. Among the drawbacks of control tissues, such as tonsil and placenta, or specially curated normal tissue (or even tumor) samples with low, medium, or high expression levels, are that they are sometimes difficult to obtain, may have an inherent biological variation that leads to variation in expression levels, and that tumors are often highly heterogeneous and may exhibit diffuse growth patterns even throughout a tissue core. Recently, the development of standardized test materials, such as cell lines, has shown promise. For example, HistoCyte Laboratories Ltd and Array Science LLC have made important break-throughs in developing standardized cell lines for testing the staining sufficiency of important predictive biomarkers.

One other source of variation in the assessment of stain quality in a routine laboratory is the human factor. The inter-observer variation is also significant for the task such as estimating the percentage of positive cells. This is particularly an issue for biomarkers that have a semi-quantitative assessment with a cut-off such as HER2 and PD-L1. It turns out that AI and image analysis can play an important role here and can be used for characterizing and quantifying staining quality. The technology has been developed and proven both for curated tissue samples and cell line materials.

To show the potential of AI and image analysis in combination with controls, we describe the following proof of concept project on which UKNEQAS and Visiopharm collaborated. As part of a pilot for the assessment of PD-L1 for triple-negative breast cancer (TNBC), UKNEQAS sent out unstained tissue slides to 16 participating pathology labs. These tissue slides contained tonsil, breast cancer tissues (low and high expression), and four cell lines from HistoCyte. After assessment by the pathologist expert panel, the slides were digitized and analyzed with AI and image analysis. Figure 13 shows examples of tonsil, one TNBC tissue, and two cell lines without and with AI-image analysis results overlaid.

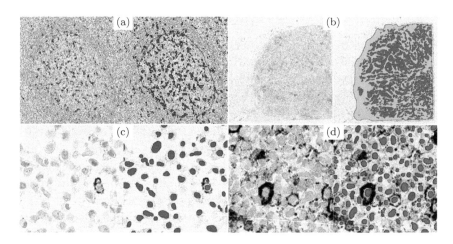

Figure 13:   Tissues and cells lines stained with PD-L1 showing (a) tonsil original and with AI results, (b) TNBC with AI results, (c) cell line (5–15%) original and with AI results, (d) cell lines (75%) original and with AI results.

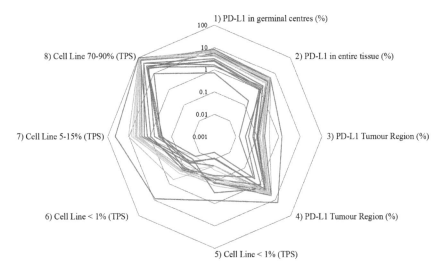

Figure 14:   Radar plot showing the percentage of positivity for different tissue types and cell lines. Shown are the results for (1) germinal centers in tonsil, (2) complete tonsil, (3) low expressing TNBC tissue, (4) high expressing TNBC tissue, and cell lines with (5) <1%, (6) <1%, (7) 5–15%, and (8) 70–90%. Each curve shows the results for a lab participating in the pilot. The blue curves are for three samples strained by a reference lab. The green, yellow, and red curves are from labs that were scored as good, borderline, and poor, respectively. Please note that the axis has a logarithmic scale.

Figure 14 shows the results of the EQA reference panel and the AI results. The radar plot shows the AI results for different tissue types and cell lines. On the axis, the percentage of positivity is shown on a logarithmic scale and each axis represents a different tissue type or cell line. The curves that are plotted in the figure are color-coded (green, yellow, and red curves) to show the EQA scores (good, borderline, and poor, respectively).

From Figure 14, it can be seen that the curves from the reference laboratory (blue) and the good quality slides (green) overlap significantly. The borderline (yellow) and poor (red) slides are further away from the reference slides. The results from the pilot suggest that there is predictive power in the use of AI for the analysis of standard controls (tonsil and TNBC tissues) and analysis of standardized cell lines. This is also shown in other publications where reference materials and AI are combined [2, 5].

Such technology can be used as decision support for EQA organizations, assisting them in achieving both standardization and scalability, and also for offering on-demand testing for pathology labs, allowing them to see trends and fluctuations in quality, and intervene to adjust protocols accordingly. In the longer term, this technology may also lend itself to a continuous and 100% on-slide quality control, which would introduce quantitative statistical techniques commonly used in the clinical pathology lab in the histology lab.

2.5. Other tissue diagnostic applications

In the text above, we have focused on how AI and image analysis applies to IHC-based tissue diagnostic biomarkers. We see that the objective and quantitative interpretation of tissue-based biomarkers is essential in the era of immune-oncology as this is key to delivering on the promise of tissue-based precision medicine and precision diagnostics. This is especially the case where the context and structure of cells and tissues are important. While in routine use, currently, most tissue biomarkers are identified using relatively straightforward monoplex techniques (such as IHC or FISH), it is expected that more complex multiplex or even hyperplex techniques may be adopted in the future due to additional information comprised within a single image and the promise of better understanding the tissue or the tumor microenvironment. Another application of AI focuses on the analysis of structural information that is visualized in H&E-stained tissue slides. These H&E slides make up most of the slides in the routine diagnostic workflow of a pathology laboratory. Slides from the organs breast, prostate, colon, lung, and skin cover the majority of the volume of slides in a typical lab. Therefore, it is no surprise that much of work has focused on applying AI to H&E-based applications. These APPs and algorithms typically aim to create efficiency in the workflow (leading to economic benefits) or to increase the quality of the diagnostic process. There are several promising examples for H&E-based applications currently being introduced in diagnostic workflows in some geographies. In Europe, several companies have developed CE-IVD marked HE-based prostate algorithms that aim to support the pathologist in detecting and grading prostate cancer. Examples for prostate, breast, and skin algorithms are brought to the market by companies such as Deep Bio Inc, Ibex Medical Analytics, Proscia, and Paige. One example of such an APP is

Figure 15: Detection of lymph node metastases in H&E-stained slides for women with breast cancer. Regions of cancer cells are recognized by a deep neural network and then outlined with a polygon in red, orange, or yellow depending on the size of the metastasis (APP 90159, Visiopharm A/S, Denmark. In US: For research use only, not for use in diagnostic procedures. In EU: CE-IVD).

shown in Figure 15 where the detection of metastatic tissue in lymph nodes is shown for breast cancer.

3. Regulatory Pathways

3.1. *Introduction*

The regulation of medical devices, throughout the world, is a vast and rapidly evolving field. Although the regulatory paths and registration processes to bring and maintain medical devices to the clinical market vary from country to country, they have common elements to the registration process including safety of the device, effective use of the device, quality design, development, and manufacturing of the device, and post-market follow-up of the device.[5] All regulatory processes are put in place to protect the patient. Regulation of medical devices in the United States is overseen by *The Food and Drug Administration* (FDA). The FDA's oversight of food and drugs began in 1906 when President Theodore Roosevelt signed the Pure Food and Drugs Act. Since then, Congress has expanded the FDA's role in protecting and promoting the development of human and veterinary drugs, biological products, medical devices and radiation-emitting products, human and animal food, and cosmetics. In 1982, the organizational units at the FDA that regulated medical devices

---

[5] *Global Medical Device Regulatory Strategy*; Peter A Takes, Susumu Nozawa, Second Edition; Chapter 2.

and radiation-emitting products merged to form the *Center for Devices and Radiological Health* (CDRH).[6] CDRH is responsible for regulating firms who manufacture, repackage, relabel, and/or import medical devices sold in the United States.

Globally, the pathology community faces a problem. Access to new, rapidly evolving pathology innovations are needed that will have a positive impact on patient care and personalized medicine. Digital pathology, both hardware and software, is no exception and the lack of regulatory authorizations are often viewed as one of the greatest barriers to adoption. The global community must come together to understand the true barriers and solve the problems using regulatory science, of which the regulatory pathway is one of the elements. Often the journey of radiology and related medical displays which started in the early 1990s, if not earlier, to go digital, is compared to the current journey of pathology to go digital. There are a lot of parallels between these two areas, but there is one big, very important difference. If it comes to tissue abnormalities seen during radiology, a pathologist is the final person in the healthcare chain to define and diagnose what type of disease a patient has and thus the input for patient management. This is contributing to the demand for clear, efficient regulatory pathways for digital pathology. In addition, general principles to develop and design algorithms based on AI are not fully clarified at this moment. This calls for a collaborative approach with FDA for each vendor, moreover, this calls for a collective collaborative community with representatives from different entities including an all-relevant stakeholders approach.

In this section, we discuss what drives the regulatory pathway, how barriers delaying adoption of clinical use of digital pathology could be overcome, and what the current, i.e., in 2021, regulatory landscape in the digital pathology field is.

3.2. *Are digital pathology technologies medical devices?*

Based on the intended use of the technology to date, digital pathology scanners, viewers, and algorithms are classified as one or more medical devices. A medical device is defined by its intended use, technology,

---

[6] https://www.fda.gov/medical-devices/overview-device-regulation/history-medical-device-regulation-oversight-united-states (accessed on March 30, 2021).

and geography. To assess if a product and its technical components is a medical device, one will have to consider the intended use and review regulations within the intended geographic region of use, to determine if it fulfills the criteria as a medical device. Once a product is assessed to be a medical device, then regulatory pathways can be defined. These pathways vary across geographies. For example, in the United States it is risk based, using a three-class system with Class I being lowest risk and Class III, the highest risk. In Europe, it is list based per IVD Directive (IVDD). Devices are divided into four groups represented into two lists, which defined whether a submission to a notified body is required. In the latter case it becomes risk based using four classes, with A being the lowest and D the highest risk per IVD Regulation (IVDR) changes in 2022. In general, the lowest risk class devices require less rigorous pre-market approval dossiers compared to highest risk class devices. In addition, different regulations can influence the composition of a device. Specifically, a device can interoperate with another medical or non-medical device. For example, interoperating devices in EU could be considered subsystems and can be regarded together as a closed system (e.g., one device). The FDA guidance *Technical Performance Assessment (TPA) of Digital Pathology Whole-Slide Imaging (WSI) Devices*[7] describes a WSI as follows: "The components in a WSI device can be grouped in two subsystems: image acquisition and image display. The image acquisition subsystem digitizes the tissue slide as a digital image file. The image display subsystem converts the digital image file into optical signals for the human reader". The TPA describes the testing at several levels including the system level, i.e., end-to-end testing. The FDA describes a system as a "series of consecutive components in the imaging chain with clearly defined, measurable input and output". As such, the WSI devices are still to be tested end to end. In other geographies, e.g., in Canada, the viewing of images itself is considered software. It can be used alone or in combination with hardware, for a medical purpose. Therefore, it can be tested and brought to market on its own. Hence, in Canada the WSI system initially was brought to market as two separate products, i.e., one to acquire an image and one to view the WSI. This provides a clear distinction between viewing *system* and viewing *software*.[8]

---

[7] https://www.fda.gov/media/90791/download (accessed on February 26, 2021).

[8] https://health-products.canada.ca/mdall-limh/ (accessed on February 26, 2021).

This approach of the imaging chain is also being followed when developing regulatory pathways for algorithms. Algorithms are being developed using images and corresponding data as input. One can understand that when the input differs, e.g., the resolution, pixels, compression level, then the algorithm could provide different outputs. As has been shown in the previous sections, algorithms could become more robust and agnostic for different inputs. Developing general principles for algorithm design and development to verify the robustness and validate the algorithmic devices in clinical settings are currently under discussions with regulatory authorities. We further outline in what follows, how these discussions can be facilitated and accelerated to provide patients access to novel quality, safe, and effective devices.

3.3. *A regulatory path forward*

As shown, regulatory pathways and the involvement of governmental affairs differ per geography. This is not only defined by regulations, but also by different needs, different opinions, and different economic drivers. These are all input to the assessment of the benefit–risk ratio and thus the risk analysis, which is a key element in the review process for registration purposes. For example, in the United States, WSI intended use for primary diagnosis was considered a Class III (highest risk) device. But for aiding a pathologist in assessing HER2, it was considered a Class II (moderate risk). While in Singapore, primary diagnosis was considered the lowest risk, and HER2 the highest risk as it was linked to a CDx. Although economics should not influence the decision-making, based on the different needs, we can safely conclude that it does play a role in risk assessment. In rural underserved areas, patients are at risk of not being diagnosed in time; digital pathology has the potential to reduce this risk and shifts the balance to benefits. Such shifts in the risk analysis also influence payor strategy,[9] the ultimate financial driver, which is directly linked to many stakeholders, such as pathology associations, healthcare administration at the federal, state, and institutional levels, manufacturers, hospital administrators, laboratory directors, and ultimately patients.

---

[9]Aligning Reimbursement for Digital Pathology with its Value by Laura Lasiter *et al.*, https://www.thejournalofprecisionmedicine.com/aligning-reimbursement-for-digital-pathology-with-its-value/ (accessed on February 26, 2021).

This highlights that digital pathology will benefit from collaborations between all these stakeholders.

US regulations are often considered the most stringent, very cumbersome and prevent early access to innovative devices that could benefit healthcare. Yet, countries around the globe are working together to harmonize medical device regulation [8]. The new EU IVD regulation (IVDR), which will go into effect in May of 2022, takes into consideration the international initiatives for medical devices, particularly, the *International Medical Devices Regulators Forum*, a voluntary group of medical device regulators.[10] The expanded IVDR will provide risk classification grouping (groups A through D) making use of performance evaluation reports, increased post-market surveillance, and a documented risk management that requires manufacturers to demonstrate that they are gathering feedback continuously and proactively. These regulation changes are a great example of this harmonization effort, where similarities to US medical device regulations are apparent. Therefore, this chapter will focus on US's Medical Device regulations, see 21CFR, which can be utilized to support further global regulatory harmonization.

3.4. *Collaborations/Collective approaches*

Regulatory and government authorities seek collaboration with each other; they want to learn from each other and work to harmonize regulations across the globe. Even so, authorities can make regulations and have the accountability and responsibility, but not always the operational understanding of how the regulation impacts day-to-day routines. Regulators and government authorities need the community key stakeholders to demonstrate clinical utility and that the medical device is safe. Therefore, medical device development, and the community at large, will benefit from consensus standards, harmonization initiatives, collaborations, and collective knowledge.

3.4.1. *Standards*

A standard is a "… document, established by consensus and approved by a recognized body, that provides, for common and repeated use, rules,

---

[10] https://www.fda.gov/regulatory-information/search-fda-guidance-documents/software-medical-device-samd-clinical-evaluation (accessed on March 30, 2021).

guidelines, or characteristics for activities or their results, aimed at the achievement of the optimum degree of order in a given context".[11] Compliance with the FDA recognized standard makes the development as well as the assessment for safety and effectiveness by an authority more efficient. Standards are known for describing processes and services which provide quality deliverables. Some examples of standards are IEEE Std 11073-20601-209 Health Informatics — Personal health device communication — Part 20601: Application profile — Optimized exchange protocol (accessed January 12, 2021) and IEEE ISO 29119-1 First edition 2013-09-01 Software and systems engineering — Software testing — Part 1: Concepts and definitions.

3.4.2. *Harmonization*

Harmonization is needed to make sure we all speak the same language and to unify regulations as much as possible. A well-known forum in the medical device industry is the *International Medical Device Regulators Forum* (IMDRF[12]), a forum of voluntary medical device regulators from around the world who have come together to build on the strong foundational work of the *Global Harmonization Task Force on Medical Devices* (GHTF[13]), and to accelerate international medical device regulatory harmonization and convergence. IMDRF defines the term *Software as a Medical Device* (SaMD) as "software intended to be used for one or more medical purposes that perform these purposes without being part of a hardware medical device". The FDA included their current thinking for developing SaMDs as well as the harmonization documents from IMDRF in their guidance document.[14] These guidelines, built on global harmonization efforts, facilitate communication and create a mutual understanding of what truly is meant between vendors and regulators; providing clarity about the design and development of the device. Specifically, it describes the required inputs, outputs, and test methods, which then has the

---

[11] https://isotc.iso.org/livelink/livelink/fetch/2000/2122/4230450/8389141/ISO_IEC_Guid e_2_2004_%28Multilingual%29_-_Standardization_and_related_activities_--_General_ vocabulary.pdf?nodeid=8387841&vernum=-2 (accessed on March 30, 2021).
[12] http://www.imdrf.org (accessed on March 30, 2021).
[13] http://www.imdrf.org/ghtf/ghtf-archives.asp (accessed on March 30, 2021).
[14] https://www.fda.gov/regulatory-information/search-fda-guidance-documents/software-medical-device-samd-clinical-evaluation (accessed on February 26, 2021).

potential to speed up the design and development of the device, facilitate review of pre-market approval submissions, and ultimately shorten time of device to market.

3.4.3. *Collaborations*

An efficient way to receive input is through collaboration. SaMDs are not new, unlike the regulation of WSI medical devices, but it is overall recognized that digital pathology technology is evolving fast and regulations cannot match the pace of the development cycle. Smart regulations are needed, which will allow for this fast pace of development to ensure timely patient access to innovation. Collaboration will only work if all involved stakeholders have a good understanding of the technology. Examples of such collaborations in digital pathology are the *Digital Pathology Association* (DPA) and the *Pathology Innovation Collaborative Community* (PIcc), which was formally known as the *Alliance for Digital Pathology.*

The DPA has proven to be effective through collaboration with the FDA, with the successful recommendation to reclassify WSI devices from a Class III (highest risk class) to a Class II device as an example. This example demonstrates why collective approaches are needed. It is in regulators' best interests to obtain input from the community and all stakeholders that are impacted by regulations. Regulators encourage collaboration; it is even a strategic initiative for the FDA's CDRH. The strategic initiative within the CDRH is to promote and support collaborative communities. Collaborative communities are regarded a continuing forum in which private- and public-sector members, which can include the FDA, work together. The community works on medical device challenges to achieve common objectives and outcomes.[15] They are convened by interested stakeholders and may exist indefinitely, produce deliverables as needed, and tackle challenges with broad impacts. In addition, FDA has installed a center of excellence, which allows the FDA to reach out to experts in the field through a non-profit organization. Collaborative communities bring together stakeholders to achieve common outcomes, solve shared

---

[15] https://www.fda.gov/about-fda/cdrh-strategic-priorities-and-updates/collaborative-communities-addressing-health-care-challenges-together#whatis (accessed on March 13, 2021).

challenges, and leverage collective opportunities.[16] As an example, PIcc was recently recognized as a collaborative community with CDRH participation.[17]

Members of the DPA and PIcc are on the frontlines. These members (1) research use and benefits of digital pathology, (2) manufacture the technology, (3) provide the greatest insights into what digital pathology entails, (4) define what the utility of it is, (5) list the risks of its use, and (6) state the impact to patient care and economics. Collaboration is a holistic approach; one in which stakeholders must keep up with current developments and routinely expand their knowledge. This can be done through research, publications, tapping into experts in the field or adjacent fields; all while feeding this valuable input back to authorities that need this context to make fair and reasonable assessments of the safe and effective use of the device. Collaboration is a win–win for all stakeholders; together we become informed to increase awareness and understanding, drive synergy, advance best practices, and can make expert judgments on devices.

3.4.4. *Collective knowledge*

Publications, white papers, and guidelines generate knowledge about how devices are used in the clinic, what the benefits, risks for misdiagnosis, and other issues with the technologies are. Examples are missed tissue during scanning, or a wrong analysis calculation, and to demonstrate how these risks can be mitigated, i.e., what are the control measures we can put in place and show the clinical utility. This knowledge will shift the benefit–risk ratio toward the benefits and allows for less strict regulatory pathways. Several landmark documents on this topic include:

- FDA's Reclassification Order for WSI devices.[18]
- A study showing that digital image analysis outperforms manual biomarker assessment in breast cancer [35].

---

[16] https://www.fda.gov/media/116467/download (accessed on February 26, 2021).

[17] https://www.businesswire.com/news/home/20210106005118/en/FDA-Participates-in-MDIC%E2%80%99s-New-Collaborative-Community-Enabling-Innovations-in-Pathology (accessed on February 26, 2021).

[18] see DEN160056, reclassification order https://www.accessdata.fda.gov/cdrh_docs/pdf16/DEN160056.pdf Table 1 (accessed on February 26, 2021).

- A study showing the high correlation of the digital imaging analysis of HER2 IHC to Her2-FISH results [10].

Other ways to build regulatory expertise and knowledge is for manufacturers to submit devices for pre-market submission authorization and share their experience, or for stakeholders to come together and create a mock submission. A mock submission is based on a hypothetical device, with hypothetical characteristics and companion information. It is also collaborative and should involve multiple stakeholders (e.g., investigators, companies, etc.).[19] Pre-submissions and mock submissions drive toward a better understanding of the technology, and generate knowledge based on regulatory science, which leverages existing regulations and can influence updates to existing and future guidance documents.

3.5. *The current regulatory landscape for AI algorithms*

At this moment, WSI devices are considered a closed system for the entire pixel pathway, which consists of scanner and viewing system. It must be tested end to end, from acquisition to display. Recently the FDA has cleared software for viewing of slides as a separate medical device, which can now be replaced within the pixel pathway through equivalence, leaving the pixel pathway unbroken. The first clearance of interoperable systems is usually tied to a specific system. In digital pathology, this was demonstrated when a cleared viewing software was swapped with one specific cleared WSI system. This was the first "swapping" of components in digital pathology, which has already been demonstrated in radiology many years ago. This type of "swapping" is the first step toward open systems and true interoperability. It opens the door to having different medical devices interoperate and exchange information with each other. It will promote the industry to develop medical devices independently, and it will enable flexibility for users. Users will not be locked to one vendor. Instead, they will be empowered to choose different vendors that support their specific workflow requirements.

---

[19] https://ncihub.org/groups/eedapstudies/wiki/DeviceAdvice/File:20190412-HTTMock Submissions.pdf (accessed on February 26, 2021).

3.5.1. *Radiology's experience using algorithms*

For pathology to draw parallel to radiology, and build upon its experience, we must first understand that the current regulatory landscape in radiology has evolved greatly since the turn of the century. When we take a closer look at that timeline, we learn the following has happened in radiology:

- Computer Aided Detection (CAD) approvals for mammography, then additional approvals followed for other intended use cases (e.g., dental, chest).
- Experience from these CADs, and other uses of AI (e.g., localization, segmentation) built experience and knowledge.
- Significant knowledge was gained through public meetings, research, guidelines, publications, and more.[20]

Radiology has the advantage of years of experience in their use of imaging and has applied this experience and knowledge to better understand clinical utility and further define special controls for risk mitigation. This enabled (1) reclassification of devices and/or (2) broader intended use cases, such as grouping single diagnoses into categories, and/or (3) allowed interoperable systems.[21]

Pathology innovation is late, relative to radiology, and has to catch up within the regulatory field. However, their learnings can enable promising outcomes for digital pathology. Radiology can be used as a precedent and facilitate pathology innovations, especially AI technology. This brings potential to speed up the regulatory paths and accelerate bringing innovation to the clinical market. Just like in radiology, the digital pathology community has a responsibility to build and share the knowledge, even when using radiology as a precedent so that we can learn, adapt, and evolve the regulatory field rapidly.

3.5.2. *Fully automated and/or continuous learning algorithms*

Modern AI technologies include both locked, static algorithms or unlocked, adaptive algorithms for pathology. Currently, only locked algorithms have received market authorization. Most of these devices are an

---

[20]https://www.fda.gov/media/135708/download (accessed on February 26, 2021).
[21]*Ibid.*

aid to the pathologist, so these algorithms are not fully automated. They are currently supportive and aid a physician in their work, but will not do the work for them or replace their role as (diagnostic) decision makers. Current harmonization guidelines, like SaMD from IMDRF, do include the option for fully automated devices. Next Generation Sequencing (NGS) is a great example of where this is already an accepted practice. The Healthcare Provider (HCP) fully relies on the NGS result, without assessing the result and even uses it for patient management.

Although there is no fully automated or continuous learning algorithm cleared by the FDA today, both industry and the scientific community are encouraged to collect Real-World Data (RWD) that can prove clinical utility and demonstrate that the data can be used for continuous learning. Yet, RWD is not always regarded as being reliable and accurate. There are risks of incompatible sources of information. For example, there was a Veteran Affairs (VA) pathologist in Arkansas who was under influence of drugs and alcohol and misdiagnosed thousands of patients.[22] If an AI algorithm is adapting to such a pathologist, or to a pathology practice with poor histology tissue preparation processes, to scanner peculiarities, or a combination of one or more poor data or error sources, it could influence the performance of AI.

Another concern of allowing continuous learning algorithms is when an algorithm that is trained on certain samples, now changes because of new samples, and starts to produce different results, even if it is an improvement. Users must become aware of this change and adopt or refute it for correct patient management. A similar concern is that only a subset of the population is used for training, but the training set should reflect the population in your intended use. This is important because authorities regulate devices to population levels. Therefore, the training and test sets should represent the intended general population and incorporate age, race, and other relevant demographics, rather than a specific small subgroup.

RWD must demonstrate how these concerns can be controlled. Usually, a small population of outliers will not influence the performance of an algorithm, and typically only lots of poor data or errors would negatively influence the continuous learning or automation of the algorithm.

---

[22] https://www.washingtonpost.com/politics/former-va-physician-charged-with-the-deaths-of-three-veterans/2019/08/20/0119ea16-c345-11e9-9986-1fb3e4397be4_story.html (accessed on February 26, 2021).

Concerns can be mitigated by building in alarms that can detect drifts or deviations from normal. This all boils down to the question "How do you create a control and report back that when an AI learns, what was true initially is still true?" When working with adaptive or continuous learning algorithms, we must establish processes for monitoring, build in controls that can measure AI drift, and incorporate representative RWD that use clinical outcomes, which is the ultimate ground truth, to ensure robust AI algorithm performance. This data, and the resulting AI algorithms, will be rich in knowledge and eventually could also provide new signatures to better diagnose patients and provide better treatment.

3.5.3. *Risk assessment for AI*

As part of the regulatory process, specific to the device in question, other risks must be identified that could impact the pixel pathway. With AI, the current thinking is that only locked, unmodifiable AI devices can be put on the market. To our present knowledge, neither have adaptive or continuous learning AI algorithms been cleared, nor are fully automated devices, that do not involve a human, available on the market (in 2020). All cleared algorithms are an aid to the diagnostic decision-making process. Even in radiology, which again is years ahead of digital pathology, CADs are still controlled by the radiologist. A good approach to think about risks, and how to implement risk control, is to think about significant change. In the rapidly changing world of AI, we must consider questions like:

- How would the device performance change, once there is an unintended or intended modification made to the device?
- How can we control unintended performance changes?
- At what point does an AI-based device have a significant change?
- When does this change become reportable/when should it become reportable?
- How should it be reported?

The AI position paper from FDA also discusses the principle of a Predetermined Change Control Plan.[23] The SaMD Pre-Specifications

---

[23] https://www.fda.gov/media/122535/download (accessed on February 26, 2021).

(SPS) describe what aspects the manufacturer intends to change through learning, and the Algorithm Change Protocol (ACP) explains how the algorithm will learn and change while remaining safe and effective. This indicates that it should be considered and taken into account to pre-specify upfront how your device is going to change and to document this, for example, in the specifications and/or intended use. It should also be specified how to test this change and what changes are acceptable. Regarding these change modifications, one also has to think about the accountability and responsibility of authorities since they are responsible for deciding whether devices can be used safely. They need to know at which point an algorithm has a significant change, and when and what to report. As a result of this, AI companies should invest time in their change control process:

- Define how to change the algorithm.
- What to change.
- How to test the change.

This extra knowledge and increasing familiarity about device performance methods, the leverage of regulations could include but is not limited to: Technical arguments, how can we reuse cases (while still separating training and testing sets), or re-image cases, and ultimately use smaller sample sizes, that is fewer cases and fewer readers, and move on to only standalone performance and eventually maybe even to no statistical hypotheses testing but just some user comparison testing.[24]

This could also open doors for device supplements, such as expanding more efficiently the indications or opening the devices and splitting them into separate devices allowing interoperability.

3.5.4. *Additional considerations*

Regulations were initially developed to safeguard and protect the public health. People have instinctively a drive to develop more guidelines and regulations due to bad experiences, and concerns about safety arise. Geographies implement their own regulations for this type of concerns,

---

[24]Presented at European Congress of pathology, Bilbao, Spain, September 9, 2018 by Brandon D. Gallas.

for example, Japan does not allow that the cloud is hosted outside Japan, EU follows GDPR, and US, HIPAA. We should strive for common sense and a holistic approach.

Allowing clinicians to speak with engineers so the designers will understand the users and can develop the device for its intended purpose, accommodating user needs, is the key to understanding the benefits and risks, allowing to build in risk control mitigations, resulting in quality devices. Such a holistic approach benefits all involved stakeholders in contrast to always wanting to regulate more or drawing up more guidance documents [23].

3.6. *Conclusion*

It is clear that digital health and related innovation provides clear advantages for patients. Even though the current regulatory landscape for AI devices has not been fully clarified, collaborative communities, including vendors are using all measures for designing and developing quality devices. We have discussed that general principles applied in radiology can also be used in digital pathology, moreover, using good machine learning practices and regulations, based on risk, could drive the path to bringing robust digital pathology devices to the clinical market.

In this trajectory, it is key that representatives of all entities collaborate to define guidelines, general principles, and standards, including their applicability. So that patients will have timely access to precision pathology.

References

[1] Althammer, S., Tan, T. H., Spitzmüller, A., Rognoni, L., Wiestler, T., Herz, T., Widmaier, M., Rebelatto, M. C., Kaplon, H., Damotte, D., Alifano, M., Hammond, S. A., Dieu-Nosjean, M. C., Ranade, K., Schmidt, G., Higgs, B. W., and Steele, K. E. (2019). Automated image analysis of NSCLC biopsies to predict response to anti-PD-L1 therapy. *J. Immunother. Cancer* 7(1): 121, doi: 10.1186/s40425-019-0589-x. PMID: 31060602; PMCID: PMC6501300.

[2] Aung, T. N., Acs, B., Warrell, J., Bai, Y., Gaule, P., Martinez-Morilla, S., Vathiotis, I., Shafi, S., Moutafi, M., Gerstein, M., Freiberg, B., Fulton, R., and Rimm, D. L. (2021). A new tool for technical standardization of the

Ki67 immunohistochemical assay. *Mod. Pathol.*, doi: 10.1038/s41379-021-00745-6. Epub ahead of print. PMID: 33536573.

[3] Baak, J. *et. al.* (2021). Visiopharm user group meeting.

[4] Bacus, J. V. and Bacus, J. W. (2000). U.S. Patent No. 6,101,265. Washington, DC: U.S. Patent and Trademark Office.

[5] Brügmann et al. (2014). ESP Abstracts 2014. *Virchows Arch.* 465: 1–379, https://doi.org/10.1007/s00428-014-1618-2.

[6] Brügmann, A., Eld, M., Lelkaitis, G., Nielsen, S., Grunkin, M., Hansen, J. D., Foged, N. T., and Vyberg, M. (2012). Digital image analysis of membrane connectivity is a robust measure of HER2 immunostains. *Breast Cancer Res. Treat.* 132(1): 41–49, https://doi.org/10.1007/s10549-011-1514-2.

[7] Dobson, L., Conway, C., Hanley, A., Johnson, A., Costello, S., O'Grady, A., Connolly, Y., Magee, H., O'Shea, D., Jeffers, M., and Kay, E. (2010). Image analysis as an adjunct to manual HER-2 immunohistochemical review: A diagnostic tool to standardize interpretation. *Histopathology* 57(1): 27–38, https://doi.org/10.1111/j.1365-2559.2010.03577.x.

[8] Garcia-Rojo, M., De Mena, D., Muriel-Cueto, P., Atienza-Cuevas, L., Dominguez-Gomez, M., and Bueno, G. (2019). New European union regulations related to whole slide image scanners and image analysis software. *J. Pathol. Inform.* 10: 2.

[9] Grunkin, M., Raundahl, J., and Foged, N. T. (2011). Practical considerations of image analysis and quantification of signal transduction IHC staining. In *Signal Transduction Immunohistochemistry*, Humana Press, pp. 143–154.

[10] Hartage, R., Li, A. C., Hammond, S., and Parwani, A. V. (2020). A validation study of human epidermal growth factor receptor 2 immunohistochemistry digital imaging analysis and its correlation with human epidermal growth factor receptor 2 fluorescence *in situ* hybridization results in breast carcinoma. *J. Pathol. Inform.* 11: 2, https://doi.org/10.4103/jpi.jpi_52_19.

[11] Heo, Y. J., Lee, T., Byeon, S. J., Kim, E. J., Shin, H. C., Kim, B., Kang, S. Y., Ha, S. Y., and Kim, K.M. (2021). Digital image analysis in pathologist-selected regions of interest predicts survival more accurately than whole-slide analysis: A direct comparison study in 153 gastric carcinomas. *J. Pathol. Clin. Res.* 7(1): 42–51, doi: 10.1002/cjp2.179.

[12] Hida, A. I., Omanovic, D., Pedersen, L., Oshiro, Y., Ogura, T., Nomura, T., Kurebayashi, J., Kanomata, N., and Moriya, T. (2020). Automated assessment of Ki-67 in breast cancer: The utility of digital image analysis using virtual triple staining and whole slide imaging. *Histopathology* 77(3): 471–480, doi: 10.1111/his.14140.

[13] Hirsch, F. R., McElhinny, A., Stanforth, D., Ranger-Moore, J., Jansson, M., Kulangara, K., Richardson, W., Towne, P., Hanks, D., Vennapusa, B., Mistry, A., Kalamegham, R., Averbuch, S., Novotny, J., Rubin, E.,

Emancipator, K., McCaffery, I., Williams, J. A., Walker, J., Longshore, J., Tsao, M. S., and Kerr, K. M. (2017). PD-L1 immunohistochemistry assays for lung cancer: Results from phase 1 of the blueprint PD-L1 IHC assay comparison project. *J. Thorac. Oncol.* 12(2): 208–222, doi: 10.1016/j.jtho.2016.11.2228.

[14]   Holten-Rossing, H. and Klingberg, H. (2019). AI deep learning tumor detection directly on ER, PR and KI67 IHC slides yields a single slide automated workflow with high concordance to manual scoring, doi: 10.13140/RG.2.2.11913.39522.

[15]   Holten-Rossing, H., Møller Talman, M. L., Kristensson, M., and Vainer, B. (2015). Optimizing HER2 assessment in breast cancer: Application of automated image analysis. *Breast Cancer Res. Treat.* 152(2): 367–375, doi: 10.1007/s10549-015-3475-3.

[16]   Kapil, A., Meier, A., Zuraw, A., Steele, K. E., Rebelatto, M. C., Schmidt, G., and Brieu, N. (2018). Deep semi supervised generative learning for automated tumor proportion scoring on NSCLC tissue needle biopsies. *Sci. Rep.* 8(1): 17343, doi: 10.1038/s41598-018-35501-5.

[17]   Koopman, T., Buikema, H. J., Hollema, H., de Bock, G. H., and van der Vegt, B. (2019). What is the added value of digital image analysis of HER2 immunohistochemistry in breast cancer in clinical practice? A study with multiple platforms. *Histopathology* 74(6): 917–924, doi: 10.1111/his.13812.

[18]   Koopman, T., de Bock, G. H., Buikema, H. J., Smits, M. M., Louwen, M., Hage, M., Imholz, A. L. T., and van der Vegt, B. (2018). Digital image analysis of HER2 immunohistochemistry in gastric- and oesophageal ade-nocarcinoma: A validation study on biopsies and surgical specimens. *Histopathology* 72(2): 191–200, doi: 10.1111/his.13322.

[19]   Koopman, T., Buikema, H. J., Hollema, H., de Bock, G. H., and van der Vegt, B. (2018). Digital image analysis of Ki67 proliferation index in breast cancer using virtual dual staining on whole tissue sections: Clinical validation and inter-platform agreement. *Breast Cancer Res. Treat.* 169(1): 33–42, https://doi.org/10.1007/s10549-018-4669-2.

[20]   Krueger, S., Gianani, R., Hirsch, B., Pieterse, S., Aeffner, F., and Young, D. (2016). Abstract 2225: Image analysis-based PD-L1 companion and com-plementary diagnostics. *Cancer Res.* 76(14 Supplement): 2225, doi: 10.1158/1538-7445.AM2016-2225.

[21]   Laurinaviciene, A., Dasevicius, D., Ostapenko, V., Jarmalaite, S., Lazutka, J., and Laurinavicius, A. (2011). Membrane connectivity estimated by digital image analysis of HER2 immunohistochemistry is concordant with visual scoring and fluorescence *in situ* hybridization results: Algorithm evaluation on breast cancer tissue microarrays. *Diagn. Pathol.* 6: 87, doi: 10.1186/1746-1596-6-87.

[22] Li, A. C., Zhao, J., Zhao, C., Ma, Z., Hartage, R., Zhang, Y., Li, X., and Parwani, A. V. (2020). Quantitative digital imaging analysis of HER2 immunohistochemistry predicts the response to anti-HER2 neoadjuvant chemotherapy in HER2-positive breast carcinoma. *Breast Cancer Res. Treatment.* 180(2): 321–329, https://doi.org/10.1007/s10549-020-05546-0.

[23] Marble, H. D., Huang, R., Dudgeon, S. N., Lowe, A., Herrmann, M. D., Blakely, S., Leavitt, M. O., Isaacs, M., Hanna, M. G., Sharma, A., Veetil, J., Goldberg, P., Schmid, J. H., Lasiter, L., Gallas, B. D., Abels, E., and Lennerz, J. K. (2020). A regulatory science initiative to harmonize and standardize digital pathology and machine learning processes to speed up clinical innovation to patients. *J. Pathol. Inform.* 11: 22, doi: 10.4103/jpi.jpi_27_20.

[24] Mukhopadhyay, S., Feldman, M. D., Abels, E., Ashfaq, R., Beltaifa, S., Cacciabeve, N. G., Cathro, H. P., Cheng, L., Cooper, K., Dickey, G. E., Gill, R. M., Heaton, R. P., Jr, Kerstens, R., Lindberg, G. M., Malhotra, R. K., Mandell, J. W., Manlucu, E. D., Mills, A. M., Mills, S. E., Moskaluk, C. A., and Taylor, C. R. (2018). Whole slide imaging versus microscopy for primary diagnosis in surgical pathology: A multicenter blinded randomized noninferiority study of 1992 cases (pivotal study). *Am. J. Surg. Pathol.* 42(1): 39–52, https://doi.org/10.1097/PAS.0000000000000948.

[25] Munari, E., Rossi, G., Zamboni, G., Lunardi, G., Marconi, M., Sommaggio, M., Netto, G. J., Hoque, M. O., Brunelli, M., Martignoni, G., Haffner, M. C., Moretta, F., Pegoraro, M. C., Cavazza, A., Samogin, G., Furlan, V., Mariotti, F. R., Vacca, P., Moretta, L., and Bogina, G. (2018). PD-L1 assays 22C3 and SP263 are not interchangeable in non-small cell lung cancer when considering clinically relevant cutoffs: An interclone evaluation by differently trained pathologists. *Am. J. Surg. Pathol.* 42(10): 1384–1389, doi: 10.1097/PAS.0000000000001105.

[26] Nielsen, S. L., Nielsen, S., and Vyberg, M. (2017). Digital image analysis of HER2 immunostained gastric and gastroesophageal junction adenocarcinomas. *Appl. Immunohistochem. Mol. Morphol.: AIMM,* 25(5): 320–328, https://doi.org/10.1097/PAI.0000000000000463.

[27] Parra, E. R., Behrens, C., Rodriguez-Canales, J., Lin, H., Mino, B., Blando, J., Zhang, J., Gibbons, D. L., Heymach, J. V., Sepesi, B., Swisher, S. G., Weissferdt, A., Kalhor, N., Izzo, J., Kadara, H., Moran, C., Lee, J. J., and Wistuba, I. I. (2016). Image analysis-based assessment of PD-L1 and tumor-associated immune cells density supports distinct intratumoural microenvironment groups in non-small cell lung carcinoma patients. *Clin. Cancer Res.* 22(24): 6278–6289, doi: 10.1158/1078-0432.CCR-15-2443. Epub 2016 Jun 1.

[28] Polley, M. Y. C., Leung, S. C., Gao, D., Mastropasqua, M. G., Zabaglo, L. A., Bartlett, J., and Nielsen, T. O. (2015). An international study to increase concordance in Ki67 scoring. *Mod. Pathol.,* 778–886.

[29] Polley, M. Y., Leung, S. C., McShane, L. M., Gao, D., Hugh, J. C., Mastropasqua, M. G., Viale, G., Zabaglo, L. A., Penault-Llorca, F., Bartlett, J. M., Gown, A. M., Symmans, W. F., Piper, T., Mehl, E., Enos, R. A., Hayes, D. F., Dowsett, M., Nielsen, T. O., and International Ki67 in Breast Cancer Working Group of the Breast International Group and North American Breast Cancer Group (2013). An international Ki67 reproducibility study. *J. Natl. Cancer Inst.* 105(24): 1897–1906, https://doi.org/10.1093/jnci/djt306.

[30] Robertson, S., Acs, B., Lippert, M., and Hartman, J. (2020). Prognostic potential of automated Ki67 evaluation in breast cancer: Different hot spot definitions versus true global score. *Breast Cancer Res. Treat.* 183(1): 161–175, doi: 10.1007/s10549-020-05752-w.

[31] Røge, R., Riber-Hansen, R., Nielsen, S., and Vyberg, M. (2016). Proliferation assessment in breast carcinomas using digital image analysis based on virtual Ki67/cytokeratin double staining. *Breast Cancer Res. Treat.* 158(1): 11–19, doi: 10.1007/s10549-016-3852-6. Epub 2016 Jun 9.

[32] Røge, R., Nielsen, S., Riber-Hansen, R., and Vyberg, M. (2021). Ki-67 proliferation index in breast cancer as a function of assessment method: A NordiQC experience. *Appl. Immunohistochem. Mol. Morphol.* 29(2): 99–104, doi: 10.1097/PAI.0000000000000846.

[33] Slamon, D. J., Godolphin, W., Jones, L. A., Holt, J. A., Wong, S. G., Keith, D. E., Levin, W. J., Stuart, S. G., Udove, J., and Ullrich, A. (1989). Studies of the HER-2/neu proto-oncogene in human breast and ovarian cancer. *Science (New York, N.Y.)* 244(4905): 707–712, https://doi.org/10.1126/science.2470152.

[34] Stålhammar, G., Robertson, S., Wedlund, L., Lippert, M., Rantalainen, M., Bergh, J., and Hartman, J. (May 2018). Digital image analysis of Ki67 in hot spots is superior to both manual Ki67 and mitotic counts in breast cancer. *Histopathology* 72(6): 974–989, doi: 10.1111/his.13452. Epub 2018 Feb 14.

[35] Stålhammar, G., Fuentes Martinez, N., Lippert, M., Tobin, N. P., Mølholm, I., Kis, L., Rosin, G., Rantalainen, M., Pedersen, L., Bergh, J., Grunkin, M., and Hartman, J. (2016). Digital image analysis outperforms manual biomarker assessment in breast cancer. *Mod. Pathol.* 29(4): 318–329.

[36] Tsao, M. S., Kerr, K. M., Kockx, M., Beasley, M. B., Borczuk, A. C., Botling, J., Bubendorf, L., Chirieac, L., Chen, G., Chou, T. Y., Chung, J. H., Dacic, S., Lantuejoul, S., Mino-Kenudson, M., Moreira, A. L., Nicholson, A. G., Noguchi, M., Pelosi, G., Poleri, C., Russell, P. A., Sauter, J., Thunnissen, E., Wistuba, I., Yu, H., Wynes, M. W., Pintilie, M., Yatabe, Y., and Hirsch, F. R. (2018). PD-L1 immunohistochemistry comparability study in real-life clinical samples: Results of blueprint phase 2 project. *J. Thorac. Oncol.* 13(9): 1302–1311, doi: 10.1016/j.jtho.2018.05.013.

[37] Varga, Z., Diebold, J., Dommann-Scherrer, C., Frick, H., Kaup, D., Noske, A., Obermann, E., Ohlschlegel, C., Padberg, B., Rakozy, C., Sancho Oliver, S., Schobinger-Clement, S., Schreiber-Facklam, H., Singer, G., Tapia, C., Wagner, U., Mastropasqua, M. G., Viale, G., and Lehr, H. A. (2012). How reliable is Ki67 immunohistochemistry in grade 2 breast carcinomas? A QA study of the Swiss Working Group of Breast- and Gynecopathologists. *PloS One* 7(5): e37379, https://doi.org/10.1371/journal. pone.0037379.

[38] Vyberg, M., Nielsen, S., Røge, R., Sheppard, B., Ranger-Moore, J., Walk, E., Gartemann, J., Rohr, U. P., and Teichgräber, V. (2015). Immunohistochemical expression of HER2 in breast cancer: Socioeconomic impact of inaccurate tests. *BMC Health Serv. Res.* 15: 352, doi: 10.1186/ s12913-015-1018-6.

[39] Vyberg, M. and Nielsen, S. (2016). Proficiency testing in immunohistochemistry — experiences from Nordic Immunohistochemical Quality Control (NordiQC). *Virchows Arch.* 468(1): 19–29, doi: 10.1007/s00428-015-1829-1.

[40] Vyberg, M. (2019). A commentary: Quality assurance in immunohistochemistry. *Appl. Immunohistochem. Mol. Morphol.* 27(5): 327–328, doi: 10.1097/PAI.0000000000000771.

[41] Wessel Lindberg, A. S., Conradsen, K., Larsen, R., Friis Lippert, M., Røge, R., Vyberg, M. (2017). Quantitative tumor heterogeneity assessment on a nuclear population basis. *Cytometry A* 91(6): 574–584, doi: 10.1002/ cyto.a.23047.

[42] Wilbur, D. C., Brachtel, E. F., Gilbertson, J. R., Jones, N. C., Vallone, J. G., and Krishnamurthy, S. (2015). Whole slide imaging for human epidermal growth factor receptor 2 immunohistochemistry interpretation: Accuracy, precision, and reproducibility studies for digital manual and paired glass slide manual interpretation. *J. pathol. Inform.* 6: 22, https://doi.org/ 10.4103/2153-3539.157788.

[43] Wolff, A. C., Hammond, M. E., Hicks, D. G., Dowsett, M., McShane, L. M., Allison, K. H., Allred, D. C., Bartlett, J. M., Bilous, M., Fitzgibbons, P., Hanna, W., Jenkins, R. B., Mangu, P. B., Paik, S., Perez, E. A., Press, M. F., Spears, P. A., Vance, G. H., Viale, G., Hayes, D. F., and College of American Pathologists. (2013). Recommendations for human epidermal growth factor receptor 2 testing in breast cancer: American Society of Clinical Oncology/College of American Pathologists clinical practice guideline update. *J. Clin. Oncol.: Official J. Am. Soc. Clin. Oncol.* 31(31): 3997–4013, https://doi.org/10.1200/JCO.2013.50.9984.

[44] Wolff, A. C., Hammond, M. E., Schwartz, J. N., Hagerty, K. L., Allred, D. C., Cote, R. J., Dowsett, M., Fitzgibbons, P. L., Hanna, W. M., Langer, A., McShane, L. M., Paik, S., Pegram, M. D., Perez, E. A., Press, M. F.,

Rhodes, A., Sturgeon, C., Taube, S. E., Tubbs, R., Vance, G. H., and College of American Pathologists. (2007). American Society of Clinical Oncology/College of American Pathologists guideline recommendations for human epidermal growth factor receptor 2 testing in breast cancer. *J. Clin. Oncol.: Official J. Am. Soc. Clin. Oncol.* 25(1): 118–145, https://doi.org/10.1200/JCO.2006.09.2775.

[45]   Wolff, A. C., Hammond, M., Allison, K. H., Harvey, B. E., Mangu, P. B., Bartlett, J., Bilous, M., Ellis, I. O., Fitzgibbons, P., Hanna, W., Jenkins, R. B., Press, M. F., Spears, P. A., Vance, G. H., Viale, G., McShane, L. M., and Dowsett, M. (2018). Human epidermal growth factor receptor 2 testing in breast cancer: American Society of Clinical Oncology/College of American Pathologists clinical practice guideline focused update. *J. Clin. Oncol.: Official J. Am. Soc. Clin. Oncol.* 36(20): 2105–2122, https://doi.org/10.1200/JCO.2018.77.8738.

Chapter 9

Tissue Cartography for Colorectal Cancer

Volker Bruns[*,‡], Michaela Benz[*,§], and Carol Geppert[†,¶]

[*]*Fraunhofer IIS, Medical Image Processing Research Group, Erlangen, Germany*

[†]*Department of Pathology, University Hospital Erlangen, Friedrich-Alexander-Universität Erlangen-Nürnberg, Erlangen, Germany*

[‡]*volker.bruns@iis.fraunhofer.de*
[§]*michaela.benz@iis.fraunhofer.de*
[¶]*carol.geppert@uk-erlangen.de*

1. Introduction

The use of artificial intelligence (AI) has been called the third revolution in pathology, the previous ones being the introduction of immunohisto-chemistry (IHC) in the 1980s and the introduction of molecular pathology in the 2000 [4]. While at the end of 2020, quite a few whole-slide scanners have already been FDA or CE-IVD certified, the *Philips IntelliSite* solution being the first one in 2017, only few AI-based digital image analysis (DIA) solutions have been FDA or CE-IVD approved to this date. More applications are likely to come. In research and academia, however, the field of *Computational Pathology* has gained a large momentum in the past years and new breakthroughs are published on a weekly basis.

The *Medical Image Processing* group of Erlangen-based Fraunhofer IIS has been active in this field for many years now. In close cooperation with the University Hospital Erlangen, one focus has been to develop solutions for assisting in the examination of colorectal cancer specimen.

This chapter will start by listing the building blocks available for analyzing whole-slide images and then continue with presenting applications and their medical use cases, summarizing our recent primary publications on fast whole-slide-cartography [2] as well as achieving robustness through domain-specific data augmentation [3] and few-shot-learning with Prototypical Networks [1].

2. Building Blocks

On a global level, the predominant objectives of employing machine learning or AI-based algorithms are to (i) detect, (ii) classify, or (iii) segment. Further, more exotic applications are virtual staining, super-resolution, or the generation of synthetic "fake" images. Recently, applications such as survival prediction and genomic prediction have also come into focus [8].

A typical example of detection tasks is to detect and count positively stained cells in an IHC-staining, e.g., cells positive for the Ki-67 proliferation marker in the context of breast cancer. Other examples are to detect tumor infiltrating lymphocytes (TILs) or count HER2 and CEP17 gene amplifications inside a number of cells during a HER2/neu FISH analysis in order to compute the average HER2 amplification ratio and derive from it the decision whether HER2 is over-expressed, in which case the HER2 blocker Trastuzumab is a therapy that is likely to help this particular patient. What these tasks have in common is that the mere presence and location of an event such as a positively stained cell in an image is required. The event does not have to be classified nor is it relevant to find its exact contour.

Classification is the task of assigning a class label from a set of predefined classes to an entire image. However, the label is not necessarily assigned to a particular location within the image. An example here is tissue classification or whole-slide cartography. A WSI is divided into overlapping or non-overlapping patches and then each patch is assigned a tissue class label such as "mucosa", "adipose", "immune infiltration", etc. Convolutional Neural Networks (CNN) are oftentimes used for this task and the size of a patch typically matches the size of the CNN's input layer and frequently ranges from 200–300 pixels in width and height.

Segmentation is the task of partitioning an image into several coherent objects. In the context of GI-pathology, an example is gland segmentation. Here, the contours of glands — also referred to as crypts — that occur in the mucosa are detected in a pixel-accurate fashion. Optionally, the inner contours can also be segmented, yielding the lumen in the case of glands. If it is required to distinguish overlapping objects or objects of different classes, then a semantic segmentation is required, which combines both segmentation and classification. An example of semantic segmentation where classification is not required is the segmentation of overlapping nuclei in the DAPI channel during FISH HER2/neu scoring. Here, it is very important to accurately distinguish overlapping cells, but from prior knowledge, it is already clear that each object must belong to the same class "cell nucleus". Oppositely, when attempting to segment and label cells in a brightfield-image, it is not known what types of cells are segmented and so an additional classification step is required.

3. Whole-Slide Cartography for Colon Histology

Whole-Slide Cartography is the semantic segmentation of an entire slide, or a Region-of-Interest thereof, into tissue classes. In a joint effort between Fraunhofer IIS and University Hospital Erlangen, a cohort of over 150 H&E stained colon resections has been digitized and hand-annotated with seven tissue classes: tumor cells, muscle tissue, connective and adipose tissue, mucosa, necrosis, inflammation, and mucus.

3.1. *Medical applications*

The resulting tissue map not only serves as an orientation for the pathologist, but also builds the foundation for a range of subsequent analyses. This chapter will lay out some potential medical applications, where a tissue map as produced by a whole-slide-cartography-analysis is useful or a pre-requirement.

3.1.1. *Presence or absence of tumor*

The presence or absence of areas labeled by the AI-based system as "tumor cells" is of high relevance. This information can be used to guide a pathologist's attention to specific areas within the specimen. An alternative use case is to utilize it as a quality assurance system that automatically executes a second look in order to ensure the pathologist has not missed perhaps a small cancerous spot in a large whole-slide. For this application, it is important to create a high-resolution, accurate cartography with high sensitivity for tumor detection and robust false positive suppression.

3.1.2. *Location of tumor area*

In molecular pathology, a tissue sample that contains a large portion of tumor cells is analyzed, for instance with a PCR. The measurements are only reliable if the tissue sample indeed contained a sufficient amount of tumor cells. This can be ensured by auto-detecting the tumor area, in particular regions with active tumor cells (and not just stroma, necrosis, mucus), and then automatically extracting one or multiple tissue cores from within this area.

The same rationale applies to extracting tissue cores for tissue micro arrays (TMAs).

3.1.3. *Tumor composition*

The tumor composition describes the ratios of different tissue types inside the tumor area. An adeno carcinoma, the most common type of intestinal cancer, originates in the mucosa from epithelial cells. Just like a healthy mucosa comprises glands and stroma in between, a tumor contains tumor cells and tumor stroma. Frequently, necrotic areas comprising dead cells as well as mucus are encountered. For instance, if a tumor contains a significant portion of mucus, it is denoted as a *mucinous tumor*. The tumor composition has been shown in the literature to be of prognostic relevance. Specifically, a high intra-tumoral stroma percentage correlates with a significantly poorer prognosis [5]. The stroma ratio also typically grows with a higher T and N class in the *pTNM* classification system. A possible reason is that stroma-rich tumors are capable of utilizing the host tissue for their growth and are therefore more aggressive.

3.1.4. *Invasive margin*

The *tumor invasion front* (also denoted as the *invasive margin*) is a margin alongside the border between the tumor and its adjacent healthy tissue. The invasion front has a constant width, the majority of which stretches into the healthy tissue, but by a small degree, it also extends into the tumor area. When a map produced by whole-slide cartography is available, it is a straightforward task to generate the tumor invasion front. By analyzing whether this front is rather smooth, curvy, or jagged, this already describes how the tumor invades the healthy tissue, specifically whether it has an infiltrative growth pattern. Koelzer *et al.* distinguish between (i) an *infiltrative* tumor border configuration, which correlates with a poorer prognosis and is frequently associated with genetic alterations such as a BRAF mutation, and (ii) a *pushing* tumor border, which correlates with a lower risk for nodal or distant metastases [6].

Additionally, locating the invasion front is a pre-requirement for calculating the tumor budding score. Here, the pathologist is required to examine the entire invasive margin and locate the hot-spot that contains the most tumor buds in a 20-fold high power field. This tedious task can be very well supported by automated tumor bud detection based on image processing and deep learning methods. When the invasive margin is known, it is also possible to automatically distinguish peri-tumoral buds (inside the invasive margin) from intra-tumoral buds (buds inside the tumor area) [9].

Probably the most important benefit of being able to locate the tumor invasion front, however, is its role in cancer research related to characterizing the *Tumor Micro Environment* (TME). Here, researchers are particularly interested in uncovering interactions between various types of immune cells in the invasive margin and the tumor itself with the goal of gaining a better understanding of how the tumor infiltrates its surroundings and deriving individual, personalized risk stratifications.

3.1.5. *Tumor invasion depth*

The depth of invasion is a significant risk factor. It is represented by the *T* in the well-known and widely used *pTNM* classification system. Table 1 shows an overview of the *T* stages in colorectal carcinoma.

When a whole-slide cartography map that distinguishes all the relevant tissue classes is available, one might be tempted to assume that

Table 1:   Description of "T" stage of pTNM staging.

Stage	Definition
*T*0	No tumor
*T*is	Restricted to mucosa (carcinoma *in situ*)
	Can be subdivided into T1m1/T1m2/T1m3 for invasion depth into top, center, bottom third of the mucosa
*T*1	Infiltration of Submucosa
	Can be subdivided into T1sm1/T1sm2/T1sm3 for invasion depth into top, center, bottom third of the submucosa
*T*2	Infiltration of Muscularis Propria
	Can be subdivided into pT2a and pT2b for infiltration of the inner circumferential layer or the outer longitudinal layer
*T*3	Infiltration of Subserosa, adipose tissue
*T*4	Perforation of peritoneum or infiltration of other organs

deriving the invasion depth would be straightforward. One simply has to analyze which healthy tissue classes are located adjacent to each tumor-labeled area. However, this task is more involved. Not all the classes and subdivisions can be inferred based merely on classifying an individual image patch. For instance, both the submucosa and subserosa contain similar looking patches of connective tissue. In addition, the subdivision of a mucosal or submucosal infiltration (T1m*, T1sm*) is based on the infiltration depth within that particular tissue layer, which requires a post-processing step that analyses the annotated whole-slide image globally. Furthermore, it is challenging to distinguish in the resulting tissue map whether the tumor has actually infiltrated an adjacent tissue layer or is merely touching it or pushing it away. Finally, especially for identifying a *pT*4 class tumor, the available tissue sections are not always suited, as the peritoneum is not always included in the sample.

3.2. *Creating a dataset*

Any image-processing algorithm, regardless whether it uses AI or not, needs to be evaluated and for this, an annotated test set that represents the ground truth is required. In addition, in the case of supervised learning, an extensive annotated training set is also required.

Table 2: Distribution of invasion depth in cohort.

Healthy	Adenoma	$pT1$	$pT2$	$pT3$	$pT4$
45.3%	2.5%	6.9%	23.9%	21.4%	0%

3.2.1. *Cohort*

The tissue cartography neural network was trained and evaluated on a cohort of over 150 colon resection samples from the University Hospital Erlangen. The resections were taken from the tumor center as well as the tumor margin. We split the set into disjoint subsets for training, validation, and testing. Additionally, we set aside a small test set for tuning the parameters of the clustering preprocessing introduced later. Roughly half of the samples show an infiltrative adenocarcinoma, the majority of those with an infiltration depth of either $pT2$ or $pT3$ (see Table 2).

3.2.2. *Selecting the classes*

We selected to train the network to distinguish between the seven tissue classes listed in Table 3.

The reason that connective and adipose tissues have been combined into a single class is that the border between these two adjacently located or even intermixed tissue patches is too time-consuming. With our current ground truth annotations, the confusion rate between these two classes was found to be too high and so we combined these classes until the annotations could be refined. The medical endpoint where the differentiation between submucosa and subserosa might be required or at least useful is the determination of the tumor's depth of invasion.

Clearly, further tissue types could be detected. Adding a new class is mostly a matter of refining the ground truth database. Additionally, it might come at the cost of a diminished overall accuracy, when the new class is not clearly distinguishable from the existing classes. Further tissue classes that should be beneficial are:

- blood and lymph vessels and, in particular, micro-vessels
- separation of connective tissue and adipose tissue
- separation in submucosa and subserosa
- separation of muscles into longitudinal and circular muscles

Table 3:   List of tissue classes the neural network was trained to distinguish.

1	Tumor cells	Regions with high amount of malignant epithelial cells Relevant for detecting presence and size of tumor and for computing tumor composition
2	Muscle tissue	Comprises mainly image patches from both the longitudinal and circular muscle, but also containing ganglia
3	Connective and adipose tissue	Large class that comprises patches from submucosa and subserosa. Initial experiments showed high degree of confusion between these two classes and so we combined them. Contains also smaller blood vessels
4	Mucosa	Comprises patches with healthy mucosa that typically contain crypts (or parts thereof)
5	Necrosis	Necrotic (dead) cells. Relevant for computing tumor composition, especially tumor-stroma ratio
6	Inflammation	Often patches from lymphoid follicles or other tissue with a significant amount of inflammatory cells
7	Mucus	Patches comprising mostly mucin. Relevant for computing tumor composition, especially tumor-stroma ratio, and for identifying mucinous tumors

- artifacts such as foreign objects, staining artifacts, air bubbles, pen marks, etc.
- events such as ulceration or erosion.

A problem already present with the selected seven classes is their imbalance (Table 4). The three classes *inflammation, mucus,* and *necrosis* account for only 4.2% of the tiles present in the dataset. This imbalance is countered by oversampling the underrepresented classes during training and by introducing data augmentation as described later. Additionally, we limited the number of tiles per slide and class to a subset of ten thousand randomly selected tiles. In other words, for the predominant tissue classes, only a subset of the patches available from our annotated dataset is actually used.

3.2.3. *Technical workflow*

Training a patch-wise tissue classification network requires a training, validation, and test dataset of labeled patches. In order to be able to establish a representative database, it is desirable that the WSIs are annotated

Table 4: Imbalance of tissue classes in training set.

Class	Tumor cells	Inflammation	Connective/ adipose tissue	Muscle tissue	Mucosa	Mucus	Necrosis	**Total**
Tiles	298,312	38,294	696,741	549,539	538,441	22,484	29,704	**2,173,515**
Percentage	14%	2%	32%	25%	25%	1%	1%	**100%**

as completely as possible. Exceptions are tissue areas that cannot be clearly assigned to one class like, e.g., slightly inflamed muscle tissue. Generally, two options exist for creating a ground truth. One option is to annotate images manually. Since this is usually a very tedious and time-consuming procedure, the other option, which is generating the ground truth automatically, is usually preferred. Examples here are to utilize special stainings, e.g., IHC, or acquisition modes, e.g., fluorescence, hyperspectral imaging, or spectroscopy. In the case of special assays, consecutive tissue sections from the same block can be scanned, where one is dyed with a standard staining like H&E and the other with a special staining, e.g., IHC. Both sections are then scanned and aligned with a whole-slide co-registration tool. Then, the ground truth contours can be computed from the special staining slide using a simple intensity thresholding operation and be projected onto the other slide. A drawback when using consecutive slides is that a cell visible in one section is typically not visible in the following section anymore due to the small distance in between two slides. This problem can be alleviated by using re-staining workflows where the same physical tissue sample is stained, scanned, washed, and re-stained.

For this project, we opted for carrying out the annotations manually. Tissue contours within the WSI were created on a Wacom pen tablet (www.wacom.com) with a pen using a whole-slide annotation software that stores the contours' points and class in a machine-readable format. However, for the training of the CNN, a collection of small image patches combined with a ground truth label is required. Therefore, the entire scan is divided into a grid of patches of a given size that depends on the input layer of the used CNN, in this case 224 × 224 pixels. Patches that are located within a manual annotation obtain the corresponding class label. If there is no or too little overlap, the patches are discarded. Patches that contain mostly white background pixels are also removed. Special care needs to be taken here to select a good threshold as the colon sections in our dataset frequently have adipose tissue at their edges, which also contain mostly white pixels. The output is a large collection of labeled patches that cover a multi-slide dataset.

The amount of patches per class in the training set is heavily imbalanced, with mucus (22,484), inflammation (38,294), and necrosis (29,704) being the smallest classes that together account for only 4.2%. We used a total of 2.2 million patches for training and capped the predominant tissue classes at 10,000 patches per class and slide.

The development of the results presented here occurred in Python using TensorFlow2, SciKit, and NumPy. For training, we use our on-site Deep Learning Cluster that comprises 34 nodes equipped with a total of $72 \times$ NVidia P100 GPUs and $16 \times$ NVidia P40 GPUs. For the tissue cartography network presented here, we selected a *ResNet50* CNN with 224 $\times$ 224 input pixel size. We employed a color augmentation method where the hematoxylin and eosin color channel are separated and then independently varied in terms of their intensity [12] by Tellez. Input patches are normalized to $\{-1.0, 1.0\}$. We trained for 15 epochs using a cross-entropy loss and the Adam optimizer, and selected a learning rate of 0.001 with a batch size of 105. The various tissue classes are represented equally within each batch. Underrepresented classes, in particular necrosis, mucus, and inflammation, are simply oversampled in each batch and so the network will not be biased toward the most frequent tissue type.

Once the training is finished, its architecture and weights are frozen and serialized into a binary file. The entire algorithm including any pre- and post-processing is then integrated into Fraunhofer IIS's virtual microscopy viewer MICAIA(r) (www.micaia.ai) that includes an interface for image analysis applications.

3.3. *Evaluation metrics*

In patch-wise classification, each patch is classified with a CNN, which produces a probability vector with one entry per class. The class corresponding to the highest probability is chosen to be the predicted label. Optionally, the highest probability can be required to lie above a chosen confidence threshold. Whenever this condition is not met, the tile is assigned to a rejection class "not sure".

After a label has been assigned, the comparison with the ground truth annotation (manually drawn tissue contours) can be carried out in different ways. Frequently, the evaluation is simply performed on a patch-level by

Table 5: Confusion matrix for two-class problem. Positive class with label A. All other classes are summed up with label B.

		Predicted label	
		A	B
Ground	A	True positive (TP)	False negative (FN)
truth label	B	False positive (FP)	True negative (TN)

$precision = \frac{TP}{TP+FP}$	$recall = \frac{TP}{TP+FN}$	$accuracy = \frac{TP+TN}{TP+FN+FP+TN}$	$F1 = \frac{2*precision*recall}{(precision+recall)}$
Equation (5) — precision	Equation (6) — recall (sensitivity)	Equation (7) — accuracy	Equation (5) — F1 score

checking for each patch in the test set whether the predicted label matches the true label. A more elaborate approach is to carry out a pixel-wise evaluation, where the labels are compared for each pixel in the WSI. The advantage is that this gives more accurate results especially at edges between tissue types. This way, a partial overlap of rectangular patches with the smooth ground truth contours as well as other prediction methods that can produce non-rectangular contours is feasible. Each compared pixel is categorized as a true positive, false positive, true negative, or false negative as shown in Table 5. Once all patches have been evaluated, the precision, recall (also referred to as sensitivity), accuracy, and F1-score can be calculated according to the equations are given in Table 5 above.

3.4. *Evaluation and visualization of tile-based approach*

The results can be visualized by assigning a color to each tissue class and colorizing each tile according to its predicted label. Adjacent tiles of the same class can be grouped into a single object, which can be represented by its contour. Optionally, the fill color can be set to be partially transparent, so that both the tissue's texture and the classification result can be seen.

When showing only a single tissue class, an alternative way to visualize the results is a heatmap.[1] Here, the probability derived by the CNN for

---

[1] Video of Fraunhofer IIS microscopy viewer with tissue cartography heatmap: https://youtu.be/bA_nrxbHQNI.

the selected class is mapped to a color, where lower probabilities are represented by colder colors (blue) and higher probabilities by warmer colors (red). In order to show the heatmap only in relevant (probable) regions, the heatmap can be restricted to patches with a probability above a given threshold or to patches where the visualized tissue class is the most probable class. In order to make the heatmap appear smoother as is common for weather forecast graphics, a smoothening filter, e.g., a Gauss filter, can be used to post-process the heatmap.

The achieved accuracy on a test set comprising 29 slides which the described tile-based approach is 93.8%. The slides were thereby divided into non-overlapping tiles. A drawback in both accuracy and visual appearance is that the tiles covering $50 \times 50 \ \mu m^2$ each, which corresponds to 224×224 pixels, can only produce a checkerboard-like result with edgy contours that do not resemble the underlying biological structures. This fact limits the accuracy that can be achieved. This limitation could in theory be alleviated by using overlapping tiles, but this would increase the analysis' computational complexity even further. In fact, with around 13 minutes per whole-slide, the analysis using non-overlapping tiles already takes very long on a modern computer using a NVIDIA GeForce GTX 1060 GPU with 1280 processing cores.

4. Fast Whole-Slide Cartography

The tile-based approach described in the previous chapter has two major drawbacks. The limited accuracy due to the edgy contours and more importantly the high computational cost motivated our work on an alternative analysis framework, titled here *Fast Whole-Slide Cartography* and under review for publication in [2]. The previously introduced tile-based classification or, in fact, any other tile-based analysis, is embedded in this framework, but preceded by pre- and post-processing steps.

4.1. Superpixel-based approach

The core of this approach is to employ a pre-processing step that clusters the whole-slide image into so-called superpixels. Ultimately, each superpixel will be assigned a tissue class label and so choosing the parameters that influence the clustering is critical. All superpixels' sizes are approximately equal. Given the goal of speeding up the overall runtime, the clustering algorithm should not add too much overhead. In our evaluation, the Simple Linear Iterative Clustering (SLIC) algorithm outperformed the tested alternatives [7]. It starts off with a regular grid of square cells and then iteratively refines the contours based on the surrounding pixel values.

Once the image has been clustered into superpixels, the tile-based approach is carried out extracting non-overlapping tiles and classifying them individually. Patches located at the intersection of two or more superpixels require special treatment. We discard patches that do not lie with at least 50% of their size within one superpixel. Afterwards, each superpixel is assigned a label that is inferred from the set of classified tiles co-located with the superpixel. Depending on the size and shape of a superpixel, a single superpixel can contain many tiles. Given that the tiles within a superpixel are not only located in the same vicinity but also have similar color, we hypothesized that it is not necessary to classify each and every tile within the superpixel just to devise a single overall superpixel label. Instead, we hoped that it would suffice to select a random subset of tiles and infer the overall label from this subset. This way, the overall runtime would be decreased significantly. Aside from the overhead for executing the clustering algorithm, the time required for classification increases linearly with the number of classified tiles. Put reversely, the fewer the tiles that are classified per superpixel, the faster the overall runtime, but the higher the risk that a superpixel will be misclassified.

In order to save time, we chose to execute the SLIC algorithm not on the full native resolution, but on a version downscaled by a factor of 16, i.e., at a resolution of 3.54 μm per pixel. Additionally, we utilize prior knowledge about the image by transforming the RGB image into a color space where the hematoxylin and eosin components are unmixed into two separate channels [10].

4.1.1. *Experimental results*

In fact, the hypothesis that it is not required to classify each single tile within a superpixel without heavily affecting the accuracy could be confirmed. A detailed description of the results is given in [2]. The main findings are summarized in the following. The clustering configuration has been evaluated and fine-tuned on a small disjoint test set comprising eight whole-slides. Eventually, we experimentally found that an average superpixel size of 3,600 pixels (at 3.54 μm per pixel), e.g., a square superpixel, would cover 0.2×0.2 mm, and a maximum number of 10 classified tiles per superpixel seems to be a good trade-off between accuracy and speed-up. On average, superpixels of this size contain 20 tiles, with a standard deviation of 6.6. Only very few superpixels fit more than 40 tiles.

While the standard patch-based approach takes on average 13 ± 5 minutes and yields an accuracy of 93.84%, the superpixel-based approach leads to an average reduction in computation time of 41%. It is obvious that the time saving stems from the fact that only fewer tiles are classified with the CNN. This by far outweighs the overhead for the pre-processing steps of separating the H&E channels and computing the clusters.

Despite the significant speed-up, the accuracy is also increased by almost two percentage points to 95.7%. Doubling the number of classified tiles per superpixel from 10 to 20 increases the accuracy only by 0.2% while increasing the runtime significantly by over a third.

Increasing the superpixel size while leaving the number of classified tiles constant further decreases the runtime, but diminishes the accuracy. However, the impact on the accuracy varies by tissue class. The accuracy for identifying the more finely grained tissue classes that are also underrepresented in the dataset, namely *inflammation*, *necrosis*, and *mucus*, suffers more heavily from larger superpixel sizes than the coarser tissue types, namely *connective and adipose tissue* or *muscle tissue*. Consequently, the average superpixel size presents a configurable trade-off between overall speed-up and accuracy of the "small" tissue classes.

A closer look at the classification results shows that the most frequent confusion appears between necrosis, or mucus, and tumor, where both are misclassified to be tumor. Both of these types of misclassification already exist in the standard tile-based approach and so a likely explanation is that the necrotic and mucinous areas are too small to fill entire tiles. However, the misclassification rate is increased by the superpixel-based approach, in the case of necrosis significantly from 14.1% to 23.4%. Here, in addition to a tile-level misclassification, a superpixel-level misclassification is the second source of error. Even when a tile is correctly predicted, it may be part of a superpixel dominated by tiles of other classes. In this case, the superpixel boundary was chosen incorrectly to begin with. Other sources of error are the manual annotations themselves: annotated tumor cell regions frequently contain small necrotic areas which might introduce a bias in the training of the CNN.

Table 6:   Comparison of runtimes and accuracies.

	Superpixel area	Max. classified tiles per superpixel	Accuracy in %	Runtime in s/slide	Runtime in %
Tile-based approach	/	/	93.84	766	100
With pre-clustering, without random tile sampling (all tiles are classified)	0,045 mm²	/	95.95	855	112
With pre-clustering and random tile sampling	0,045 mm²	10	95.67	448	58
With pre-clustering and random tile sampling	0,125 mm²	10	94.87	203	27

4.2. *Uncertainty metric*

The introduction of a rejection class can aid in increasing the specificity. The rationale here is that in cases where the neural network is not very certain about its decision, i.e., it has a low confidence, it might be better to indicate this to the user by not assigning a class at all rather than assigning a potentially wrong class. Certainly, this is not an option in all applications, but in the case of tissue cartography a "not sure"-class can be acceptable. The drawback to this procedure is that it will also decrease the sensitivity in cases where a low-confidence prediction is rejected but was actually true. Thus, it is crucial to define or measure the level of confidence as accurately as possible with the goal of boosting the specificity without at the same time overly sacrificing the sensitivity.

In a standard tile-based approach, the straightforward way of defining the confidence is to examine the distribution of the class probability vector emitted by the neural network's output layer. In our case, a softmax function is used, so that the sum of all class probabilities accumulates to one. To give an example, when presented a patch right from the transition between healthy mucosa and a tumor, the neural network might produce a value of 0.5 for *mucosa*, 0.4 for *tumor*, and smaller values accumulating to 0.1 for the remaining classes. A similar probability distribution might occur if presented a patch from a low-grade adenoma. It is clear that since the decision between two classes is almost 50/50, the confidence of this decision is not high. Consequently, we might introduce a threshold of, for instance, 0.7, stating that only predictions where the most probable class has a softmax output of 0.7 or higher are considered and all others are assigned to the rejection class. It was shown that although the direct correspondence between softmax output and confidence is low, softmax output is still useful for detecting misclassified and out-of-distribution examples [13]. A superior method, which is widely used, is to require a low entropy of the estimated class distribution, which yields the maximum value for a constant distribution where all classes are equally probable.

In the case of superpixels, however, we can base the uncertainty not only on the probability distribution of a single tile, but on all the classified tiles that correspond to the same superpixel. We defined five different uncertainty metrics and compared them is no longer contained in current version of that publication. Three metrics are based directly on the softmax outputs, as described above. Here, we simply average the softmax vectors of all classified tiles of a superpixel. In the first case the softmax entropy is used and in the second case the maximum value is used directly as a confidence measure. A third alternative option is to evaluate the difference between the most and second most probable classes. Two more metrics are not based directly on the softmax output, but instead on the distribution of votes for each class among all classified tiles in a superpixel. Again, one way is to use the number of votes for the most probable class and an alternative is to evaluate the difference between the most and second most probable classes. In all metrics, a higher value indicates a higher confidence.

4.2.1. *Experimental results*

We evaluated for all five metrics how the number of true and false predictions changes in dependency of the chosen confidence threshold. A higher confidence threshold will lead to more superpixels being assigned to the rejection class. A good metric will initially reject mostly false predictions and only few true predictions. The question is which metric does the best job at separating true from false predictions and what threshold is a good trade-off? All metrics show the desired behavior that they discard proportionally more false than true predictions. When decreasing the amount of false predictions to 20%, the amount of true positive predictions has been reduced only to around 90% by all metrics. However, it is important to note that since the overall accuracy of the classification is already very high to begin with and there are many more true than false predictions, in absolute numbers all metrics discard more true than false predictions.

Aside from a good separation, a desirable property of a confidence metric would be that it can be acutely controlled and this would be the case if the value range between a confidence threshold where almost no superpixels are rejected and the threshold where almost all superpixels are rejected is very large, ideally one. However, our measurements show that the range is very narrow for the softmax entropy-based variant, where this range stretches only from 0.0 to 0.05. Likewise, for the maximum softmax metric this range stretches only from 0.15 to 0.3, and from 0.0 to 0.2 for the metric where the difference between the highest and second highest average softmax value is considered. If these narrow value ranges stay constant for other datasets, this is less problematic otherwise a suitable standardization has to be investigated. Further experiments will have to be carried out in this direction. In terms of the threshold value range, on this dataset, the two metrics based on the class votes in all classified tiles of a superpixel perform best. The metric based on the delta between the classes with the highest and second highest vote count seems to be slightly superior in terms of distinguishing true from false predictions. The disadvantage here is that this metric takes only discrete values depending on the number of classified tiles.

We evaluated the class-wise response to a confidence threshold and found that the smaller tissue classes, namely inflammation, mucus, and necrosis, are more strongly affected. Sampling a minimum required confidence of 0.45, already 5% of the superpixels predicted to contain necrosis and 3% of inflammation as well as mucus superpixels were weeded out due to a low confidence, while for mucosa this ratio is only at 0.5%. Clearly, the confidence for mucosa predictions is higher on average. The reason for this could be that these three classes, due to their occurrence in smaller areas, are more often not accurately detected in superpixel segmentation and are mixed with other tissue types. In contrast, the mucosa usually occurs in large contiguous regions.

4.3. *Evaluation of medical applications*

This chapter will present the results of experiments carried out to evaluate the potential medical applications listed above.

4.3.1. *Tumor area*

The tumor area is defined by fusing tumor cells and enclosed stroma, necrosis, and mucus. We evaluated the tumor area using a separate test set of $N = 18$ samples. For each sample, two consecutive slides are available, one with a standard H&E staining, the other with the IHC epithelial marker pan-cytokeratin (AE1/AE3). The latter is not yet used for evaluating the tumor area, but will be used later for assessing the tumor composition, in particular the active tumor cells. We annotated the tumor contours manually in the H&E and IHC slides. The tumor area is not directly available from the predicted tissue cartography results as a tumor consists of multiple tissue types: tumor cells, mucin, necrosis, and tumor stroma. Consequently, deviations in the predicted tumor area can stem from multiple sources.

(1) Automatically detected superpixel contours are likely to follow the biological structures more closely than the manual annotations.
(2) Superpixels can be misclassified.
(3) Superpixels of different tissue classes can be combined incorrectly into a tumor area.

The accuracy is assessed using the metrics *Intersection over Union* (IoU) and the *Dice* coefficient. In summary, the accuracy is very satisfactory. The annotated tumor area differs on average by only 6% with an IoU of 89.4% and a Dice coefficient of 94.3%. Most of the misclassified areas are located at the tumor boundary (Figure 1).

However, the test set contains three outliers with larger discrepancies. At closer inspection, the problem is not that patches or superpixels have been misclassified. Instead, the rule whether or not necrosis or mucus

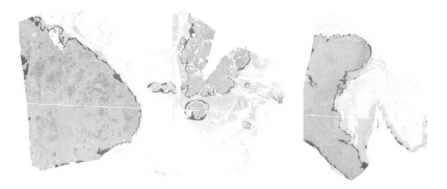

Figure 1:   Visual comparison of predicted tumor area with ground truth.

adjacent to tumor-cells regions are counted toward the tumor area or not is ambiguous. In one case, an untypically large necrotic area is flanked by tumor areas on both sides. Our tumor area algorithm joined these three large areas into a single large tumor, but the pathologist in this case did not count the necrosis toward the tumor area and instead counted both adjacent tumors as separate entities. In another case, necrotic areas adjacent to the lumen had been counted as part of the tumor, but had been excluded by the pathologist. In a third case, adenomas that had not been found to be malignant by the pathologist have been misclassified as tumor by the tissue analysis.

This preliminary evaluation on a relatively small dataset of only 18 slides shows promising results and indicates that a tissue cartography map is a very helpful starting base for locating and measuring the exact tumor area, which in turn allows for the computation of the tumor composition, in particular the tumor–stroma ratio. However, a more sophisticated superpixel fusion algorithm is required to capture the intricacies of the tumor biology. This is subject to further research.

4.3.2. *Invasive margin*

After the tumor area has been identified and segmented, it is straightforward to define and locate the tumor invasion front, which is located at the boundary between the tumor and the adjacent healthy tissue except mucosa.

As explained above, being able to automatically locate the invasive margin is helpful in at least three applications. Firstly, especially in the context of immune oncology, researchers investigate the interaction between the immune cells and the tumor in the tumor microenvironment (TME). Secondly, an application in routine diagnostics is the scoring of tumor budding, which involves finding the budding hot-spot along the invasive margin. Thirdly, the invasive margin by itself has prognostic value. It can be classified as either an *infiltrative tumor border* or a *pushing tumor border*, the former correlating with a poorer prognosis [6].

Two pathologists qualitatively evaluated the generated margins of all slides of dataset B using a point-based grading system from 1 (very good) to 5 (not sufficient). With an average score of 1.6, the subjective evaluation indicates that the computed whole-slide cartography provides a solid basis for inferring the tumor margin. We did not yet attempt to classify the tumor margin in "infiltrative" vs. "pushing". A straightforward approach could be to compute the margin's roundness or to count the number of curves and normalize by its length.

4.3.3. *Tumor composition*

Quantifying the tumor composition, in particular the tumor–stroma ratio, is of high prognostic relevance. It has been show in the literature that a high stroma ratio correlates with a poorer prognosis, the hypothesis being that the particular tumor seems to be able to utilize healthy host tissue in its growth [5].

We found that the granularity of the manual ground truth annotation for the active tumor area was not sufficient to accurately measure the tumor composition. Instead, the ground truth is extracted from an adjacent serial section stained with the epithelial marker pan-cytokeratin AE1/AE3 (PCK). Special care must be taken, since both healthy and malignant cells express this marker. We use a simple algorithm to accurately detect all contours encircling positive cells or cell clusters in the entire IHC WSI. It unmixes the RGB color image into separate DAB and Hematoxylin images and then binarizes the DAB image with a carefully selected threshold that captures all positive cells (strong brown) and at the same time rejects necrotic areas (light brown). This mask is then intersected with the manually annotated tumor areas. The resulting mask is then assumed to cover only the vital tumor cells inside the tumor area and serves as a ground truth.

This evaluation showed that the active tumor area predicted by the superpixel approach deviates on average by over 30% from the ground truth extracted from the PCK IHC stain. The purely patch-based approach in comparison performs only slightly better.

A closer look revealed that both methods systematically overestimate the vital tumor cell area. Clearly, using superpixels or even single tiles as the smallest unit of measurement is insufficient to capture the finely grained structures within a tumor. As a reminder, superpixels have an irregular shape and were configured to cover at most an area of 0.2×0.2 mm, if they were square. Tiles are always square and cover an area of $50 \times 50 \ \mu$m. For this reason, we tested a third alternative. We employed a separate neural network that was pre-trained for the task of segmenting glands in healthy mucosa. After all, the tumor cells in an adenocarcinoma are epithelial cells arranged in crypts. With this gland segmentation network, the average deviation is decreased to 11%, which is a lot better, but still leaves room for improvement. Inspecting the results visually, we noted that the deviations were largest for samples where the crypt architecture was more heavily distorted. For this reason, we evaluated the accuracy in the test set separated by tumor grade. Indeed, for slides with grade 1 or 2, the average deviation is below 4% while the average deviation for slides with grade 3 is again over 20%. For the superpixel and patch-based methods, the tumor grade does not seem to have an influence on the accuracy. We conclude that analyzing the tumor area with a separate network trained particularly to segment tumor cells both in low- and high-grade tumors should be a viable solution. The existing gland segmentation network should provide a good starting point. It could be extended to segment high-grade tumors as well using transfer learning. This would entail that a dataset comprising high-ground tumors needs to be manually annotated and used to re-train either the entire existing gland segmentation network or possibly freeze all except the last few layers. Again, this is subject to future research.

Although of lower clinical relevance, we also evaluated how well our proposed methods perform for predicting the areas of either necrotic or mucinous areas within the tumor. Both of these tissue types were captured quite accurately. The absolute deviation between the predicted and true relative necrotic area ratio of the tumor is under 2%. While all tumors in the test set contained at least some areas with necrosis, only 5 of the 18 slides contained mucin. Constraining the evaluation to these five slides, the deviation is a little over 0.5%. Averaged over the whole test set it decreases to around 0.2%. We did not observe a notable difference in accuracy between the superpixel and purely patch-based methods. We conclude that no extra analysis step is required to determine the area of these two tissue classes.

5. Coping with Real-Life Heterogeneous Data

The results presented in the previous chapter were measured on a single dataset. Despite the fact that the dataset had been properly split into disjoint training, test, and validation subsets, the fact remains that all slides stem from the same hospital, have been prepared in the same lab, and were digitized with the same scanner.

This in no way guarantees that the analysis will yield a similarly high accuracy when presented with slides prepared at a different lab or scanned with a different scanner. With the goal of deploying the system anywhere in the world, it is crucial to tackle this problem and make the classifier more robust to the types or heterogeneity it may encounter.

The sources of variance are three-fold:

(1) Wet-lab: Different labs may employ different chemicals, different mixture ratios, and different protocols. Frequently, pathologists claim that they can identify from which lab and geographic region their sample had been prepared merely by looking at the staining color.

(2) Hardware: The most critical components of a tissue scanner that influence the resulting image appearance are the camera sensor, the microscope objective, and the lighting.

(3) Color settings/Post-processing: The scanner vendor might choose to employ a specific exposure duration or color gains during image acquisition or boost the contrast or saturation in a post-processing step.

We conducted experiments limited to variations introduced by using different scanners and will report the results, originally published in [3], in this chapter. While we observed significant variations, it might well be that the variations present in a multi-centric dataset are even larger. Applying the tissue cartography trained on colorectal resections to other entities can also be regarded as an increase in variance. Making the classification more robust as well as finding and pushing the limits of its applicability will be the focus of our work in the coming months and years.

5.1. *Multi-scanner dataset*

The first step is to validate if the problem of a diminished accuracy for WSIs from other scanners really exists and if it does, quantify it. To this end, a multi-scanner database has been created.

The starting point was the single-center single-scanner dataset used in the previous chapters, which was scanned with a *3DHISTECH MIDI* at the Institute of Pathology of the University Hospital Erlangen, Nuremberg. The training set comprises over 2 million patches from 92 slides and the validation set around 700,000 patches from a disjoint set of 30 slides. We then digitized a disjoint set of 39 glass slides with five additional scanners of different vendors, creating a multi-scanner dataset

Table 7: Multi-scanner dataset.

Scanner	Resolution [μm/px]	Number of test patches	Number of training-patches for fine-tuning
Number of WSIs	—	30 per scanner	9 per scanner
3DHISTECH MIDI	0.22	1,381,316	40,230
Fraunhofer iSTIX (manual scanning)	0.17	2,123,364	49,005
Fraunhofer SCube	0.27	857,511	38,528
PreciPoint M8	0.35	514,397	35,524
Hamamatsu NanoZoomer S210[1]	0.22	1,424,716	—
Hamamatsu NanoZoomer S360[1]	0.23	1,298,056	—

[1] Scans courtesy of Hamamatsu Germany.

of 234 WSIs (Table 7). Four of these scanners are regular automated whole-slide scanners by *Fraunhofer IIS* (*SCube* prototype), *Hamamatsu* (two *NanoZoomer* models[2]), and *PreciPoint* (the *M8*), while the fifth scanner is the manual WSI scanning solution *iSTIX* developed by *Fraunhofer IIS*. *iSTIX* is a software that analyses in real-time the video feed from any non-motorized camera-equipped microscope and stitches together video frames into a panoramic whole-slide image that is assembled on the computer monitor as the operator moves the microscope stage back and forth.

In an additional experiment, we tested if the accuracy can be increased further by not limiting the training data to *3DHISTECH* scans and employing augmentation, but by additionally fine-tuning the pre-trained network on a small amount of image patches from the *3DHISTECH*, *PreciPoint* and *Fraunhofer* scanners. We deliberately chose not to include patches from the *Hamamatsu* scanners in order to evaluate if this fine-tuned network generalizes to unseen scanners (Table 7, right-most column).

The scans of each glass slide were then co-aligned using a rigid global registration. Once the scans are aligned to each other, the ground truth annotations manually created on the original WSIs digitized with the *3DHISTECH* scanner can be transferred to the other scans as well. Now that vector annotations are available for all scans, tiles can be extracted and labeled, producing a per-scanner set of labeled patches. Baseline accuracies are then measured by classifying the tiles from the new scanners using the network that was trained on the tiles extracted from the original *3DHISTECH* dataset.

5.2. *Strategy*

The baseline experiments verified the problem that the accuracy drops when the network is applied to scans from an unknown scanner. The extent is even larger than was anticipated. A baseline experiment, where the network trained only on the *3DHISTECH* scans was used to analyze WSIs produced by the other scanners without implementing any counter-measures, showed that the accuracy drops from 94% to below 40% for all except the *Fraunhofer SCube* scanner (see Table 8).

Essentially, two strategies can be devised for making the network more robust to the heterogeneity encountered in a real-world setting. The first strategy is to attempt to transform whole-slide images from unknown scanners at runtime prior to analysis in an attempt to make their charac-teristics match those of the slides in the training dataset. This process is referred to as normalization. The predominant solution has been to unmix the hematoxylin and eosin channels using a color deconvolution. Then the channels' intensities are normalized and the image is transformed back into RGB. Other authors have reported that this method can produce unre-alistic results and so is prone to errors. A more recent solution is to use generative adversarial networks (GANs), which can be used to transfer the style from one image to the other. Mainstream non-medical examples where GANs were used are the web service "simpsonize me" that con-verts people from real portrait photos into characters from the TV series *The Simpsons*, or the conversion of arbitrary images into the style of Van Gogh's painting *Starry Night* or Edvard Munch's painting *The Scream*.

The other strategy is to attempt to reflect the full bandwidth of vari-ance already in the training set. This way the network will be trained to cope with the variance from the start. The literature suggests that the latter approach yields better results and so this is what this chapter will focus

Table 8: Baseline accuracies obtained when training network only on 3DHISTECH slides without color augmentation.

3DHISTECH	PreciPoint	Fraunhofer		Hamamatsu NanoZoomer	
MIDI	M8	iSTIX	SCube	S210	S360
94%	39%	29%	73%	35%	36%

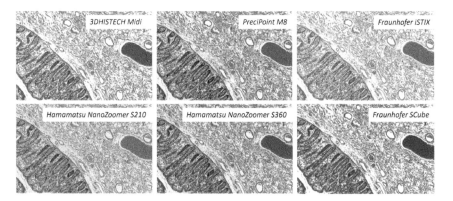

Figure 2:   Visual comparison of whole-slide images digitized with six different scanners.

on [12]. One way to assemble such a heterogeneous training set is to collect real samples sourced from multiple laboratories. An alternative way is to employ data augmentation in order to generate synthetic data. The question is what augmentation techniques to employ. To this end, it is an important first step to identify and characterize the differences between the different scans (Figure 2).

The two strategies mentioned above do not utilize scans from a new unknown scanner. An additional option is to fine-tune the existing network using a transfer learning approach. It would be the superior solution if it were possible to train a network to yield accurate results on an unseen WSI from an unknown scanner. However, if this problem cannot be solved to full satisfaction, the next best solution is to establish a workflow that allows to fine-tune an existing network with transfer learning using only a few additional annotated slides of the unseen scanner. In this case, we started off with the original CNN that was trained on 2 million patches from the *3DHISTECH* scanner with color augmentation enabled and then fine-tuned with only 40,000 patches per scanner extracted from nine WSIs from four selected scanners.

5.3. *Data augmentation*

One important difference between the scans produced by different scanners is that they differ in their resolution in terms of μm per pixel.

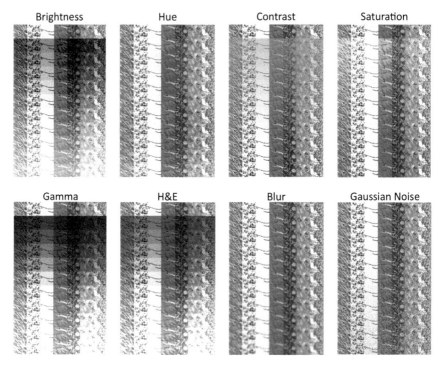

Figure 3: Domain-specific color augmentation. Top row of each of the eight blocks shows original patches from 3DHISTECH scans. Below rows are synthetically produced.

This property can be reliably determined at runtime and compensated for by scaling the patches prior to analysis to the same resolution used in the training set.

The main visual difference is the color. Therefore, we applied a set of augmentations that accept a patch from the *3DHISTECH* scans as input and produce a copy of that patch that is varied in terms of that particular color or geometric property (Figure 3). The strength of the variation is selected randomly, while the upper and lower bound of the variation can be selected. The varied color properties are hue, saturation, brightness, contrast, and gamma. Additionally, the H&E stain separation method is also utilized as an augmentation technique, where the intensity of either hematoxylin or eosin or both is varied. As geometric augmentations, we used scaling. Not included in the experiments but shown for illustrative purposes are the blur and Gaussian noise variations.

The image augmentation techniques are automatically employed as part of the training process. We defined for each augmentation type a probability and valid parameter range. During training, when a patch from the training set is requested, it will automatically be augmented. Whether or not a particular augmentation type will be applied is defined by its assigned probability. Each augmentation type is tested independently, i.e., a single patch can and will likely undergo several augmentations back to back. While the motivation for the augmentation here is primarily to introduce variance, it is also useful to compensate for the class imbalances. We ensure that per batch each of the classes is represented equally often on average and so, consequently, underrepresented classes are oversampled. When oversampling a class, it is useful to use augmentation in order to reduce the likelihood that identical patches are presented to the network.

5.4. *Experimental results*

As a network architecture, we used the *Xception* CNN [14] with an input patch size of 224×224 px. The network was trained with a cross-entropy loss and Adam optimizer with a learning rate of 0.001 and exponential decay.

The baseline accuracies had been determined by training the network only on the *3DHISTECH* slides without employing any data augmentation and then classifying the test set of the other five scanners. We did not correct for the different resolutions at runtime. One finding was that the color has a higher impact on the accuracy than the resolution, because the accuracy on slides digitized with the *Fraunhofer SCube* scanner (73%), which have a similar color but different resolution, are much higher than those for the Hamamatsu scanners (35%), which have a different color but identical resolution. A test where we did correct for the different resolutions at runtime showed that this does not make a significant difference — at least for the variations in resolution present in this multi-scanner dataset. The color properties are indeed very different. We computed and compared the average hue and saturation values over a subset of slides per scanner and observed strong differences. Especially, the *iSTIX* scans have a significantly lower average hue and saturation. When we scanned the slides with *iSTIX*, we chose to adjust the camera's color settings to resemble what we observed through the microscope's binoculars rather than the *3DHISTECH* whole-slides.

We were then interested in finding out what accuracy could be achieved simply by training the network instead on grayscale versions of the original *3DHISTECH* slides. Surprisingly, this simple trick already increased the accuracy significantly to around 85% for all automatic scanners. The drawback is that the maximally achievable accuracy, represented by the measured results on the native *3DHISTECH* dataset, was decreased by three percentage points to 90.8%. This is a serious limitation as the robustness shall not be achieved with a significant diminishment of the native-scanner accuracy.

We then set out to determine the impact of various augmentation methods individually. The zoom augmentation yields only less than 2% difference from the baseline experiment. This confirms the initial finding that the variations in resolution at least within the present range do not have a strong influence on the accuracy. From the color augmentation types, the hue and H&E augmentation have by far the strongest positive influence. The H&E augmentation increases the accuracy of the *PreciPoint* and Hamamatsu scanners from around 35% (baseline) to over 65%, while decreasing the accuracy on the original dataset only by 0.7 percentage point. The hue augmentation performs even better and yields an accuracy of 85%.

The next question is whether the gains achieved by the individual augmentation types add up — at least partially — when used in combination. Indeed, the accuracies for the automatic scanners increases slightly to values between 88% and 90% when using hue, saturation, and H&E augmentation in combination (Table 9, fourth block of rows). This means the individual benefits do not fully add up, but it is nonetheless advantageous to employ these three augmentations in conjunction.

Finally, we were interested in finding out if the accuracy can be further increased by fine-tuning the existing network trained on the *3DHISTECH* scans. To this end, we continued the training on a small multi-scanner dataset that contains only around 40,000 patches per class extracted from nine slides per scanner. We only included patches from the *PreciPoint* and *Fraunhofer* scanners, and not from the *Hamamatsu* scanners. Our experiments showed that it is essential to include again patches from the original *3DHISTECH* scanner during the transfer learning process, or otherwise its accuracy will drop significantly (Table 9, bottommost block). The result is that the accuracy can indeed be increased further, but in this case only for the scanners represented in the fine-tuning

Table 9: Overview of achieved accuracies with multi-scanner dataset. For the baseline, absolute accuracies are stated, all other values state the difference to the respective scanner-baseline. The third block shows the independent benefit of the tested color augmentations. The fourth block shows the benefit when combining the most helpful augmentations. The last block uses the Hue + Saturation + HE augmentations and trains the network that was pre-trained on 2 million 3DHISTECH patches additionally on 40,000 patches extracted from 9 WSIs from each named scanner.

	3DHISTECH	PreciPoint	Fraunhofer		Hamamatsu	
	MIDI	M8	iSTIX	SCube	NZ S210	NZ S360
Baseline (abs.)	93.9	39.4	29.0	73.1	35.4	36.1
Grayscale	−3.1	45.5	39.0	15.4	51.0	52.1
Zoom	−0.8	2.4	2.0	1.9	−1.1	−1.7
Gamma	−1.8	0.4	8.0	11.7	−4.9	−3.0
HE	−0.7	28.5	11.1	10.0	28.3	31.9
Hue	−1.2	**46.7**	12.0	12.1	**46.7**	**53.5**
Saturation	−0.5	5.8	7.5	**12.5**	6.8	9.8
Brightness	−2.1	10.0	8.1	8.4	1.8	3.6
Contrast	−1.2	6.1	4.6	6.5	1.4	1.9
HE + Gamma	−1.4	30.2	14.2	15.3	30.9	36.6
Hue + Saturation + HE	−2.1	49.7	**33.1**	**16.5**	**52.6**	54.0
Hue + Sat. + HE + Contrast	−0.6	**50.0**	20.3	12.7	50.9	**55.6**
M8 fine-tuning	−32.9	54.5	22.9	2.8	31.4	41.9
iSTIX fine-tuning	−37.3	24.8	53.1	−11.7	22.2	17.2
SCube fine-tuning	−4.5	19.8	9.6	20.5	13.5	17.5
MIDI + M8 + iSTIX + SCube	−3.7	52.3	56.2	17.3	42.4	43.8

Header over data columns: Accuracy increase w.r.t. to scanner's baseline

Source: Adapted from Ref. [3].

training set. Compared to the accuracy achieved by employing hue, saturation, and H&E color augmentation alone, it could be increased by 2.6%, 23.1%, and 0.8% for the *M8, iSTIX,* and *SCube*, respectively. However, the accuracy on the *3DHISTECH* scanners drops to 90.2%, which is a problem. Also, the effect does not seem to generalize to the Hamamatsu scanners. Their accuracy drops by over 10%.

5.5. *Conclusion*

Clearly, more research is required to achieve an accuracy on new scanners that is comparable to that observed on the native scans. However, by employing the proposed color augmentations that introduce variance in hue, saturation, and the individual hematoxylin and eosin channels' intensities, the accuracy achieved on new scanners can be increased by up to 50%, yielding an absolute accuracy of almost 90% in all scanners, except *iSTIX*. The accuracy measured on the *iSTIX* scans lies constantly below the other scans. The reason may be that the manual stitching concept is more prone to quality errors. In particular, manual stitching lacks an autofocus feature as the microscope is not remote controllable and the software can only work passively. The operator is in charge of adjusting the focus when necessary, which is especially challenging for samples with an irregular surface. In future work, we will additionally experiment with introducing a blur augmentation. The cost of generalizing the model in order to make it robust against changes introduced by using a different scanner is that the accuracy of analyzing the native scans decreases. This is not surprising as the model is optimized to perform well on a much broader problem and naturally the task of classifying only slides from one and the same scanner is easier and will allow for a higher accuracy. The accuracy is diminished by 2% when using only color augmentations and by 3.7% when additionally fine-tuning with a small training set of patches scanned with other scanners. The benefit of this fine-tuning is limited for the *M8* (+2.3%) and *SCube* (+1%) but significant for *iSTIX* (+23%). It does not generalize to the unseen *Hamamatsu* scanners though, that have not been represented by the small fine-tuning training set. We hypothesize that their color characteristics lie outside of the range spanned by the other scanners. In future work, this will have to be investigated further.

In summary, these experiments have shown that a neural network cannot be expected to produce acceptable results when presented scans from other scanners. The drop in accuracy was tremendous, which underlies the need for investigating how to train robust models that can eventually be deployed anywhere in the world.

However, using color augmentation, it is possible to trade a small decrease of accuracy (−2%) on the native scans for a dramatic increase (+50%) on unseen scans and this process does not even require scans, let alone annotated scans, of the new scanners. If, however, scans of a new scanner are available, a transfer learning approach, where the pre-trained network is fine-tuned, can increase the accuracy on the new scanners further, but at the cost of another diminishment of the native accuracy (Figure 4). More experiments are required and the individual scanner datasets' color properties need to be properly characterized with the goal of predicting based on these properties whether a network will be able to generalize to scans from that scanner. Another major limitation of the experiments carried out thus far is that they only investigated the heterogeneity introduced by using other hardware, but did not evaluate the impact of using different preparation or staining protocols. To this end, this single-center multi-scanner dataset needs to be extended to a multi-center multi-scanner dataset.

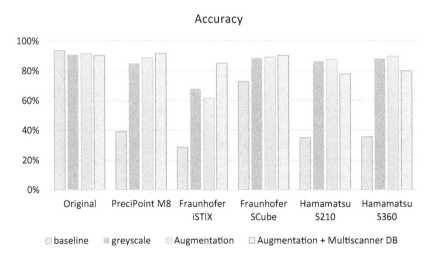

Figure 4: Comparison of benefits from color augmentations.

6. Learning by Example

The method proposed above utilizes a CNN which is trained to solve a fixed problem: to distinguish between seven tissue classes. It can do that, but only that. The input to the CNN is an image and the output is a vector with one probability per class. Adding an eighth class or using seven different classes is not possible without re-training the network. Given the sheer number of problems in histopathology, where so many different pathologies in so many different organs have to be examined and each score evaluates different morphologies particular to the specific disease at hand, it seems like a good idea to consider developing instead a workflow where new problems can be tackled more easily without having to label vast amounts of data and train a network from scratch every time. This is what motivated the concept of "Few-Shot-Learning", where a network is taught to solve a problem based on only a very limited amount of training data. The basic idea is to pre-train a network and make it "familiar" with the type of data it will encounter over the course of its lifetime, but without already restricting it to or rather specializing it on a very specific task. In a regular CNN architecture, the classification itself is typically carried out by a fully connected layer including a softmax at the end. When stripping away this layer, the CNN can instead be used to produce a feature-vector that describes the given input in a discriminative way. The number of features, i.e., the vectors dimensionality, depends on the chosen model architecture. The N-dimensional space enclosing the feature vectors is referred to as the feature space, or latent space. What each feature represents exactly is unknown. The field of explainable AI (xAI) is concerned with providing insights on how a network functions, what areas in the source image it pays attention to, or which neuron is activated in which situation. One particular method for realizing Few-Shot-Learning is the so-called *Prototypical Network* [15]. It consists of a combination of a CNN that is used to produce a low-dimensional feature vector and a dynamic classifier that is not restricted to a fixed number of classes. In fact, this method can even be used to only introduce discrete classes at runtime, e.g., by the user, as opposed to at training time by the developer. Each class is represented by a prototypical feature vector, hence the name. A simple way of defining new classes for an application is to have the user present a set of examples that belong to this class. The CNN is then used to compute a feature vector for each example and these vectors can be combined, e.g., take the average, to define the class prototype. This process is repeated for as many classes as desired. New, unknown images are then classified by computing their feature vector and calculating the distance in the feature space to all class prototypes. The image is assigned to the class with the closest distance. If the feature space were three-dimensional, each class could be thought of as representing the center of a sphere and new images can be assigned to the sphere of closest proximity.

During the training of a CNN, it is optimized by minimizing a loss function. It is up to the developer to define or select a loss function. When training a regular CNN to solve one particular problem, the loss function can simply return the error between the predicted labels and then the true labels. The *categorical cross-entropy* (CCE) function is a frequently selected loss function. This error is then minimized during training. In the context of *Few-Shot-Lea*rning, we introduce additional constraints in the loss function with the goal of teaching the CNN to position feature vectors of different classes far apart in the latent space. On the other hand, images showing the same class should be positioned in close proximity. In particular, we use the *Clustering-Oriented Representation Learning* loss, in short COREL, which does exactly this [11]. It contains an attractive component that rewards positioning similar images in close proximity in the latent space and a repulsive component that rewards increasing the distance between dissimilar images.

In the context of histopathology, *Prototypical Networks* are an exciting technology that immediately poses a set of research questions. How widely applicable is a given pre-trained CNN that is used to produce the feature vectors? If trained on H&E slides, can it also be used for classifications in IHC slides or other stainings? What about the resolution? Can it be trained on 20× scans and then be used to classify 10× or 40× scans? If pre-trained with gastro-intestinal samples, can it be used to solve problems in dermatopathology, nephropathology, or neuropathology? If the CNN network is pre-trained on a wide range of samples from different organs, scanners, resolutions, staining, will it still be able to yield competitive accuracy? And what is the penalty in accuracy anyway when compared to a network that is trained the "classic" way to solve only one particular problem? And how many samples from how many different sections are needed exactly to find a good prototype that well represents a new class — tens, hundreds, or thousands?

6.1. *Experimental setup*

We cannot answer all these questions yet, but did set out to quantify the capabilities of *Prototypical Networks*. Given the availability of the multi-scanner database presented in the previous chapter, we re-formulated the goal of achieving robustness on slides digitized with other scanners as a few-shot-learning problem. The large set of labeled patches scanned with the *3DHISTECH MIDI* scanner presents the database our *Prototypical Network* was pre-trained on. We then treated patches from a new unknown scanner as a new class represented by their own prototype. The results presented in the chapter stem from our publication in [1].

In a first experiment we compared the accuracy of a *Prototypical Network* and a "classic" CNN. Both are exclusively trained on patches from the *3DHISTECH MIDI* scanner. They yield comparable accuracies. When testing with patches from the *MIDI* scanner, the accuracy is around 90% and the other scanners' accuracy drops significantly without applying any counter-measures as was already shown in the previous chapter. We then utilized the true potential of the *Prototypical Network* and adapted the class prototypes using only 1,000 patches per class extracted from nine annotated slides each digitized with a new scanner, which is a not a lot of data compared to the over 2 million patches from 92 WSIs the CNN was pre-trained on. Indeed, the accuracies on the test sets of the new scanners now rise significantly: on the *M8* from 40% to almost 60% and on the *Hamamatsu S360* from 35% to almost 60%. It is not fair, however, to compare a *Prototypical Network* trained on scanner-specific data with a regular CNN that was only trained on the *MIDI* scans. When including the data augmentation techniques presented in the previous chapter, indeed the regular CNN's accuracy rises to somewhere between 80 and 90% on the new scanners and outperforms the *Prototypical Network*.

Not giving up on the concept of *Prototypical Networks,* we then considered that it must be a drawback that they are forced to define a class in the latent space merely by a single prototypical center point, which approximates the class distribution as a sphere in the latent space with the prototype being the center. We employ the well-known *k-means* algorithm to cluster the set of feature vectors into k clusters and so far used $k = 1$. Indeed, when describing a class by two ($k = 2$) instead of one prototype, the accuracy measured on *M8* scans rises from 60% to 67%.

Using three prototypes yields an additional small increase, leading to an accuracy of 70%. Using more than three prototypes, however, did not boost the accuracy further. These observations are consistent regardless whether we used the accuracy or the F1 score as an assessment metric.

Since data augmentation and (*Multi-*) *Prototypical Networks* are not alternative options but could be combined, we then set out to evaluate if and how much the individual benefits of both methods add up when combined. Now, data augmentation is activated when pre-training the CNN module of the *Prototypical Network* on the original *MIDI* scans. To recap, the CNN classically trained on the MIDI scans without any data-augmentation achieved an accuracy of only 40% on the M8 scans, 35% on the *Hamamatsu* scans, 32% on the iSTIX scans, and close to 70% on the *SCube* scans (which subjectively look most similar to the *MIDI* scans). The *Prototypical Network* with three prototypes per class, but without data augmentation, yields a significant average improvement of 26 percentage points compared to the classical method without data augmentation. But the accuracies benefit even more significantly when using a classic CNN and using data augmentation during training. The accuracies for the new scanners range between 83% and 87% for all new scanners except for iSTIX (60%). The *Prototypical Network* where data augmentation was addition-ally enabled during pre-training does indeed achieve the best results and improves upon these numbers. By one to four percentage points for all automated scanners yielding accuracies between 87% and 88%. For the *iSTIX* scans, the result is even further improved upon by 10%, increasing it from 60% to 70%.

In summary, *Prototypical Networks* are helpful in this scenario, but nowhere near as effective as data augmentation. However, these meth-ods can be combined and together achieve the best results, very close to the accuracy obtained when evaluating the original CNN on the MIDI scans that it was also trained on. Especially on the *iSTIX* scans, however, the result is still unsatisfactory. With *iSTIX* being a manual scanning solution most prone to artifacts such as motion blur or out-of-focus areas, we suspect that the variance introduced in the training dataset using data augmentation does not yet include the type of heterogeneity present in the *iSTIX* scans. Adapting the class prototypes using *Prototypical Network* does help a lot here, but not enough. Introducing additional data augmentation methods such as artificial blur should prove to be helpful.

6.2. Stability

One important research question is to find out how much the accuracy of a *Prototypical Network* depends on the examples that were used to compute the class prototypes. It is important to gain some insights on their behavior in this regard in order to be able to judge how reliable the results are. To this end, we ran experiments where we pre-trained a *Prototypical Network's* CNN on the *MIDI* scans (without data augmentation) and then adapted the class prototypes using only 1,000 patches per class extracted from nine scans created with a new scanner, in this example the *PreciPoint M8*. Using 1,000 patches may sound a lot, but compared to the 2.1 million patches the network was trained on, this presents only 0.3% of additional data. In terms of the extra work required to adapt a network to a new scanner, the number of slides to be annotated is the more relevant number as many patches can be extracted from a single slide. And annotating only 9 new slides seems like a reasonable effort.

We examined if it has a high influence whether the example patches used to define the class prototypes are extracted from 3, 5, 7, or 9 different scans. Even the full set of nine slides does not contain the full 1,000 patches for the classes *inflammation* and *mucus* for all scanners. Also, when restricting the number of slides further, for necrosis the amount of available patches drops below 1,000. We repeated all experiments at least 40 times with other randomly selected patches for all classes where a sufficiently large number of patches was available.

As expected the achievable accuracy drops slightly when basing the class prototypes on a smaller number of slides. In the best case of using three prototypes per class, the accuracy increases from 69% when using three slides, to 70% when using five slides, and to almost 71% when using seven slides. At the same time, the standard deviation decreases from 2.6% to 1.1%. The results hint that already a subset of five annotated new slides seems to be sufficient to define robust class prototypes.

6.3. What's next?

As has become clear, many research questions in relation to the use of Multi-Prototypical Networks in histopathology remain unanswered. Going forward, we want to further investigate the helpfulness of this technology beyond its utilization for coping with the variance introduced during preparation of digitization of samples. We are eagerly looking forward to solving truly new classification tasks that involve new tissue or cell types, different tumor entities, samples from different organs, or different stainings using a workflow where the true experts — the pathologists — can themselves seamlessly create new classifiers merely by teaching it with a few problem-specific examples.

Acknowledgments and Funding

Data, annotations, and medical consultancy are largely provided by Dr. Geppert and his colleagues at the Institute of Pathology, University Hospital Erlangen, Friedrich–Alexander Universität Erlangen–Nürnberg (FAU).

Assembling a dataset, training networks, and conducting experiments is a lot of work and requires a large team of skilled researchers and research assistants. We would like to especially thank our colleagues involved in these projects: Petr Kuritcyn, Jessica Deuschel, Daniel Firmbach, David Hartmann, Serop Baghdadlian, Jakob Dexl, Dominik Perrin, and Martin Weidenfeller.

This work, in particular the aspects related to designing robust classifiers, was supported by the Bavarian Ministry of Economic Affairs, Regional Development and Energy through the Center for Analytics — Data — Applications (ADA-Center) within the framework of "BAYERN DIGITAL II".

The work related to accelerating the analysis using a superpixel-clustering as well as aspects of explainable AI were supported by the Federal Ministry of Education and Research under the project reference number 01IS18056A TraMeExCo.

A part of the employed scanners and microscopes as well as Fraunhofer IIS's on-site Deep Learning cluster have been funded by the Federal Ministry of Education and Research under the project reference numbers 16FMD01K, 16FMD02, and 16FMD03.

References

[1] Deuschel, J., Firmbach, D., Geppert, C. I., Hartmann, A., Bruns, V., Kuritcyn, P., Dexl, J., Baghdadlian, S., Hartmann, D., Perrin, D., Wittenberg, D., and Benz, M. (2021). "Multi-Prototype Few-shot Learning in Histopathology", *Proceedings of the International Conference on Computer Vision (ICCV), CDPath workshop (accepted)*, https://openaccess.thecvf.com/content/ICCV2021W/CDPath/papers/Deuschel_Multi-Prototype_Few-Shot_Learning_in_Histopathology_ICCVW_2021_paper.pdf.

[2] Wilm, F., Benz, M., Bruns, V., Baghdadlian, S., Dexl, J., Hartmann, D., Kuritcyn, P., Weidenfeller, M., Wittenberg, T., Merkel, S., Hartmann, A., Eckstein, A., and Geppert, C. I. (2021). Fast whole-slide cartography in colon cancer histology using superpixels and CNN classification, publication pending, preprint: https://arxiv.org/abs/2106.15893.

[3] Kuritcyn, P., Geppert, C. I., Eckstein, M., Hartmann, A., Bruns, V., Dexl, J., Baghdadlian, S., Hartmann, D., Perrin, D., Wittenberg, T., and Benz, M. (2021). Robust slide cartography in colon cancer histology — Evaluation on a multi-scanner database. *BVM Proceedings*.

[4] Salto-Tellez, M. *et al.* (2019). Artificial intelligence — the third revolution in pathology. *Histopathology* 74(3): 372–376.

[5] Huijbers, A., Tollenaar, R. A., v Pelt, G. W. *et al.* (2013). The proportion of tumor-stroma as a strong prognosticator for stage II and III colon cancer patients: Validation in the VICTOR trial. *Ann. Oncol.* 24(1): 179–185, doi:10.1093/annonc/mds246.

[6] Koelzer, V. H. and Lugli, A. (2014). The tumor border configuration of colorectal cancer as a histomorphological prognostic indicator. *Front Oncol.* 4: 29. Published February 18, 2014, doi:10.3389/fonc.2014.00029.

[7] Achanta, R., Shaji, A., Smith, K., Lucchi, A., Fua, P., and Süsstrunk, S. (2010). SLIC Superpixels, EPFL Technical Report no. 149300.

[8] Srinidhi, C. L., Ciga, O., and Martel, A. L. (2021). Deep neural network models for computational histopathology: A survey. *Med. Image Anal.* 67: 101813, https://www.sciencedirect.com/science/article/abs/pii/S1361841520301778.

[9] Bergler, M., Benz, M., Rauber, D., Hartmann, D., Kötter, M., Eckstein, M., Schneider-Stock, R., Hartmann, A., Merkel, S., Bruns, V., Wittenberg, T., and Geppert, C. I. (2019). Automatic detection of tumor buds in pancytokeratin stained colorectal cancer sections by a hybrid image analysis approach. In *Digital Pathology. ECDP 2019. Lecture* Notes *in Computer Science,* Vol. 11435, Springer.

[10] Ruifrok, A. C. and Johnston, D. A. (2001).Quantification of histochemical staining by color deconvolution. *Anal. Quant. Cytol. Histol.* 23(4): 291–299.

[11] Kenyon-Dean, K., Cianflone, A., Page-Caccia, L., Rabusseau, G., Cheung, J. C. K., and Precup, D. (2018). Clustering-oriented representation learning with attractive-repulsive loss. CoRR abs/1812.07627.

[12] Tellez, D., Balkenhol, M., Otte-Hoeller, I., van de Loo, R., Vogels, R., Bult, P., Wauters, C., Vreuls, W., Mol, S., Karssemeijer, N., Litjens, G., van der Laak, J., and Ciompi, F. (2018). Whole-slide mitosis detection in H&E breast histology using PHH3 as a reference to train distilled stain-invariant convolutional networks. In *IEEE Transactions on Medical Imaging*, 37(9): 2126–2136. September 2018, doi: 10.1109/TMI.2018.2820199.

[13]   Hendrycks, D. and Gimpel, K. (2016). A baseline for detection misclassi-
       fied and out-of-distribution examples in neural networks. *International
       Conference on Learning Representations 2017*, arXiv:1610.02136 [cs.NE]

[14]   Chollet, F. (2017). Xception: Deep learning with depthwise separable con-
       volutions. *Conf. Comput. Vis. Pattern Recognit.* pp. 1800–1807.

[15]   Snell, J., Swersky, K. and Zemel R. (2017). Prototypical networks for
       few-shot learning. In *Advances in Neural Information Processing Systems*.
       pp. 4077–4087.

Chapter 10

Graph Representation Learning and Explainability in Breast Cancer Pathology: Bridging the Gap between AI and Pathology Practice

Pushpak Pati[*,†,¶¶], **Guillaume Jaume**[*,‡,‖‖],
Antonio Foncubierta-Rodriguez[§,***], **Florinda Feroce**[¶,†††],
Giosue Scognamiglio[¶,‡‡‡], **Anna Maria Anniciello**[¶,§§§],
Nadia Brancati[‖,¶¶¶], **Maria Frucci**[‖,‖‖‖], **Daniel Riccio**[**,****],
Jean-Philippe Thiran[††,††††], **Orcun Goksel**[‡‡,‡‡‡‡], and
Maria Gabrani[§§,§§§§]

[†]*IBM Research — Europe, Zürich, Switzerland;
Department of Information Technology and Electrical Engineering,
ETH Zürich, Zürich, Switzerland*

[‡]*IBM Research — Europe, Zürich, Switzerland;
Ecole Polytechnique Fédérale de Lausanne (EPFL),
Signal Processing Laboratory (LTS5), Lausanne, Switzerland*

[§]*IBM Research — Europe, Zürich, Switzerland*

---

[*]Equal contribution.

¶*National Cancer Institute-IRCCS-Fondazione Pascale, Naples, Italy*

‖*Institute for High Performance Computing and Networking — CNR, Italy*

***University of Naples, "Federico II", Italy; Institute for High Performance Computing and Networking — CNR, Italy*

††*Ecole Polytechnique Fédérale de Lausanne (EPFL), Signal Processing Laboratory (LTS5), Lausanne, Switzerland*

‡‡*Department of Information Technology, Uppsala University, Sweden; Department of Information Technology and Electrical Engineering, ETH Zürich, Zürich, Switzerland*

§§*Cognitive Healthcare and Lifesciences, IBM Research — Europe, Zürich, Switzerland*

¶*pus@zurich.ibm.com*
‖*gja@zurich.ibm.com*
****fra@zurich.ibm.com*
†††*f.feroce@istitutotumori.na.it*
‡‡‡*giosue.scognamiglio@istitutotumori.na.it*
§§§*a.anniciello@istitutotumori.na.it*
¶¶*nadia.brancati@icar.cnr.it*
‖‖*maria.frucci@cnr.it*
*****daniel.riccio@unina.it*
††††*jean-philippe.thiran@epfl.ch*
‡‡‡‡*orcun.goksel@vision.ee.ethz.ch*
§§§§*mga@zurich.ibm.com*

1. Breast Cancer Pathology in Clinical Practice

According to the global cancer statistics released on December 14th, 2020, by the International Agency for Research on Cancer (IARC), the estimated new breast cancer cases in 2020 reached 2.26 million (24.5% of the total number of cancer cases for females), making female breast cancer the most commonly diagnosed cancer globally[1], for the first time. While it is the leading cause for death in women with cancer, the survival rates for breast cancer are eminently high, subject to early diagnosis and adequate treatment. A study by [1] shows that for patients diagnosed with breast cancer during 2005–2009, the five-year survival rate rose to 85% or higher in 17 countries worldwide. This has been achievable due to early diagnosis through mass screening programs or intensive early diagnostic activity. Therefore, accurate, early diagnosis and risk assessment of breast cancer subtype for a patient holds a paramount role in the treatment and survival of the patient [2, 3].

In a clinical setting, a pathologist manually inspects a tissue specimen to detect and assess breast lesions, and thereby, to estimate any cancer intrinsic subtypes of the lesions given a predefined grading system. The subtypes confer different levels of risk according to their probability of transitioning to invasive carcinoma. For instance, lesions with atypia or ductal carcinoma *in situ* (DCIS) are associated with higher risks compared to benign lesions [4, 5]. The manual inspection by a pathologist starts with the discernment of coarse morphological and topological distribution attributes of the tissue, followed by the localization and analysis of specific regions of interest (RoIs). Further assessment is confined to the RoIs to analyze the phenotype and organizational properties of cells for subtyping the tissue specimen. Although diagnostic criteria for cancer subtypes are established, the continuum of histologic features phenotyped across the diagnostic spectrum prevents having clear decision boundaries between the cancer subtypes. Thus, manual inspection is a time-consuming process with significant intra- and inter-observer variability [5–7]. For instance, a study by [5] shows that the inter-pathologist agreement can be as low as 48% for breast lesions with atypia. The aforementioned challenges in manual diagnosis and the increasing incidence rate of breast

---

[1] https://www.uicc.org/news/globocan-2020-new-cancer-data#.

cancer cases per year [8] demand the use of automated computed-aided cancer diagnostics.

1.1. *Role of AI in digital pathology*

The advent of whole-slide imaging systems, producing high-resolution giga-pixel sized whole-slide images (WSIs) from tissue specimens, and the approval from the U.S. Food and Drug Administration (FDA) for using whole-slide scanners in the pathology workflow [9] have profoundly transformed the daily practice of pathologists [10–15]. Advances in imaging technology enabled the pathologists to review and interpret digital pathology slides and motivated innovative research opportunities in computer-aided diagnostics to leverage artificial intelligence (AI) [16–18], especially machine learning (ML) and deep learning (DL), to address various computational pathology tasks. The objective of AI is to assist the pathological diagnosis in clinical settings through a sound, efficient, robust, reproducible, and objective decision support system. AI can discern and learn complex patterns from WSIs, to characterize tissue structures and thereby deliver high performance predictive analysis.

AI techniques, thus, have been leveraged to address numerous pathology tasks, such as nuclei detection [19, 20], nuclei segmentation [21, 22], nuclei classification [23, 24], gland segmentation [25, 26], tumor detection [27–29], tumor grading [30–33], tumor staging [27, 34], biomarker quantification [35–37], staining optimization [38, 39], staining normalization [40–42], survival analysis [43, 44], and molecular profiling [45]. Notably, task-specific AI techniques have been developed across various cancer types, e.g., lung [46], breast [47], prostate [48], colon [49], stomach [50], liver [51]. In particular, AI has achieved human-level performance for lymph node metastasis detection [28], detecting cancer in prostate biopsies [95], grading of prostate biopsies [33, 96], skin cancer classification [97], and classification and mutation prediction from non-small cell lung cancer [98].

1.2. *Expectations from AI for pathological adoption*

The role of AI in computational pathology is growing with the advancements in imaging techniques, curation of task-specific datasets, data storage, innovative research in AI, and efficient and powerful computational resources. The trend positively indicates that AI can be adopted in clinical practice to assist pathologists' daily routine. However, to this end, AI solutions should ideally accommodate a number of expectations, listed as follows:

- *Generalizability*: A pathologist with expertise in a specific pathology task can apply his/her learning and experience to unseen similar pathology tasks. Whereas task-specific AI solutions require their own large, annotated datasets which are expensive, cumbersome to acquire, and even infeasible for several tasks. Thus, there is a need for AI models that are transferable and that can easily adapt to new tasks by leveraging information learned on related domains.
- *Explainable and interpretable*: Today's AI solutions aim to automatically learn the mapping between input pathology data and expected pathology outcome. The automated mapping is incomprehensive and is detached from pathological reasoning. Thus, current implementations of AI in computational pathology produce "black-box" AI models. Further, the AI solutions can only be evaluated based on the performance metric on the targeted pathology task. Though AI achieves satisfactory performance on several tasks, such performance is achieved at the expense of reduced transparency, which is imperative in any medical diagnosis, prognosis, and therapeutic response. Thus, there is a need for explainable and interpretable AI-based decision-making procedures.
- *Performance*: The AI solutions achieve performance comparable to expert pathologists on several pathology tasks. However, the performance evaluations do not consider the incurred risk of misdiagnosis. For instance, misclassifying a benign lesion as malignant incurs less risk compared to a malignant lesion being misclassified as benign. Further, considering the infeasibility of training AI algorithms on all possible scenarios of a particular task, the evaluation of an AI solution's robustness becomes essential. Other valuable assessments of AI solutions include time and memory efficiency for practical realization, as well as reproducibility under training and testing scenarios. Thus, AI solutions are expected to be sound, robust, reproducible, and time and memory efficient, while providing accurate diagnosis.
- *Human-machine co-learning*: The current AI algorithms intake and process raw pathology data. Such processing does not utilize

well-established task-specific prior pathological knowledge. Further, the AI algorithms designed to process raw data are less flexible in accommodating domain knowledge, task-specific criteria, and any desired analytical procedures. Development of purely data-driven and hence pathologically detached AI solutions hamper pathologists' trust in the solutions, thereby impacting their adoption in a clinical setting. Thus, there is a need for the AI solutions to be flexible in incorporating prior pathology knowledge and to facilitate building trust with pathologists to enable human-machine co-learning.

1.3. *Limitations in clinical adoption of AI*

Given its recent groundbreaking success, AI holds the promise to significantly alter the way we diagnose and stratify cancer in pathology. DL techniques represent a milestone in this transformation, as the application of deep neural network models are already behind several breakthroughs addressing key current issues of histopathology. Among several types of deep neural networks, convolutional neural network (CNN) is the most commonly used in histology image analysis. CNN facilitates end-to-end learning through high capacity and flexible models to perform routine diagnosis of the histology images or to identify novel insights into disease. However, in view of the above-listed expectations from AI for adoption in a clinical setting, the CNNs are limited in several aspects. Primarily, a CNN processes a histology image in a patch-wise manner. It extracts representative patterns from the patches in the image and aggregates the patch-wise representations to perform image-level tasks. The employed patch-wise processing suffers from the trade-off between the resolution of operation and the acquisition of adequate context information [52, 53]. Operating at a high-resolution captures local cellular information but limits the field-of-view due to expensive computational burden, thereby limiting the access to global information concerning the tissue microenvironment. Contrastingly, operating at a low resolution hinders access to the cellular and subcellular microenvironment. To address the issue with visual context, [52] proposes a context-aware stacked CNN, [53] proposes a multi-scale late-fusion CNN, and [54] proposes a CNN using neural image compression. However, these CNN architectures, operating on fixed-size input patches, are confined to a fixed-size field-of-view. Additionally, the fixed-size inputs restrict end-to-end CNNs from adapting to the largely variable sizes of typical histology images.

Most importantly, pixel-based processing CNNs disregard the notion of pathologically comprehensible biological entities [55]. The disparity between the comprehension of a histology image by a pathologist and a CNN prevents the interpretability and explainability of the CNN's input space in clinically meaningful terms. Several explainability techniques have been developed in computational pathology, such as feature attribution [55–57], concept attribution [58], and attention-based learning [59, 60]. However, these techniques produce visually blurry and unlocalized explanations. The explanations are limited in depicting tissue composition [55] and are not expressed in pathologically understandable terminologies. The absence of the notion of biological entities also restricts the inclusion of well-established, biological, entity-level, prior pathological knowledge into a CNN's design or operation, thereby hindering the human–machine co-learning. Also, CNNs disregard the structural composition of tissues, where finer entities hierarchically form coarser entities, such as, epithelial cells organized to formulate epithelium, which further constitute to form glands, and so forth.

1.4. *Role of graphs in computational pathology*

The limitations of pixel-based processing in CNNs are addressed by shifting the analytical paradigm to entity-based processing. In the entity-based paradigm, a histology image is described as an entity-graph, where the nodes of the entity-graph represent biologically defined entities and encode the entity morphologies, and the edges of the entity-graph denote entity-to-entity interactions and encode inter-entity interactions. The entity set, entity attributes, graph topology, and inter-entity relationships in an entity-graph can be configured by leveraging task-specific prior pathological knowledge. Thus, an entity-graph representation can be made pathologically consistent to allow for human–machine co-learning. An entity-graph representation of a histology image is also memory-efficient compared to a pixel-based representation, making entity-graph representations scalable to arbitrary image sizes. Thus, these can seamlessly describe a large histology image by including a large number of nodes and edges. Given the entity-graph representation of a histology image, AI techniques operating on graph structured data, such as graph neural networks (GNN), can conduct histology image modeling.

Entity-based processing using cells as the entity set, i.e., a cell-graph (CG), has been introduced in [61], where nodes and edges of the CG depict cells and cellular interactions in a histology image. The CG representation has motivated the development of several other entity-graph-based methodologies in computational pathology [62–71]. The introduced methodologies propose various CG representations by configuring node-level encoding, graph topology, employed AI technique, and others to address several pathology tasks. Though CGs efficiently encode a cell's microenvironment, they are suboptimal in comprehensively describing a histology image. For instance, the set of cells cannot extensively capture the tissue microenvironment, i.e., the distribution of tissue regions such as necrosis, stroma, epithelium, and lumen in an image. Notably, an entity-graph comprising the set of tissue regions cannot represent the cellular interaction and local cell-level information. Therefore, it can be concluded that an entity-graph representation using a single type of entity set cannot entirely describe the tissue structure in a histology image. Consistently, an AI solution modeled on a single-level entity-graph representation results in suboptimal histology image analysis.

The entity-based processing enables the interpretability of the input space of the AI solutions to pathologists. Thus, the employment of interpretability and explainability techniques on the AI solutions can furnish pathologically comprehensible explanations. However, research on explainability and visualization using entity-graphs has been scarce. CGCNet [63] analyses the cluster assignment of nodes in a CG to group them according to their appearance and tissue types. CGExplainer [72] introduces a post-hoc graph-pruning explainability technique to identify decisive cells and interactions. Robust spatial filtering [73] utilizes an attention-based GNN and node occlusion strategy to highlight cell contributions. Although graph explanations resulting from explainability techniques are often intuitive, they are not always expressed in pathologically understandable terminologies, e.g., "How large are the important nuclei?", "How irregular are their shapes?", etc. This limitation hinders the comprehension and adoption of entity-graph explanations by pathologists.

We address the aforementioned limitations of entity-graph representations, entity-graph-based learning, and entity-graph explainability in our proposed methodology, *HistoCartography*. The methodology constructs an entity-graph representation, Hierarchical Cell-to-Tissue (HACT), consisting of multiple types of entity sets, i.e., cells and tissue regions to encode both the cell and tissue microenvironment. As the multisets of entities are inherently coupled, the depiction of the structural composition of the tissue denotes a multi-scale hierarchical entity-graph. The entity-graph encodes the morphology, intra- and inter-entity relationships of the multiset of entities. It describes a histology image from fine cell scale to coarse tissue-region scale. Interestingly, the proposed entity-graph resembles the tissue diagnostic procedures in clinical practice, where a pathologist hierarchically analyses a tissue from coarse to fine level. Subsequently, our methodology introduces a hierarchical GNN, the Hierarchical Cell-to-Tissue Network (HACT-Net), to perform the histology image analysis. HACT-Net processes the HACT graph sequentially, from low-level to high-level, to provide a fixed dimensional feature representation that encodes the morphological and topological distributions of the entities in the tissue. Upon such histological image analysis, several graph explainability techniques are employed to produce intuitive graph explanations. Along with subjective qualitative assessment, the methodology proposes a set of quantitative metrics to objectively assess and express the generated graph explanation in pathologically understandable terminologies.

2. HistoCartography

Digitized histology slides are processed by the proposed entity-based methodology, HistoCartography, to meaningfully map the tissue composition to tissue functionality. The overview of the HistoCartography methodology is presented in Figure 1. In this work, for simplicity, we presume

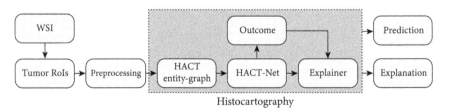

Figure 1: Overview of the proposed computer-aided diagnostic workflow by incorporating the proposed HistoCartography methodology. The workflow results in automated, task-specific, expected pathological outcome and a visually comprehensible explanation for the automatically generated prediction.

that tumor RoIs in a scanned WSI are pre-detected, and we focus the analysis on the tumor RoIs (TRoI). The methodology begins with pre-processing the detected TRoIs. Any task-specific pre-processing can be performed to normalize the histology images originating from various sites, imaging techniques, staining protocols, etc. The histology images can also be pre-processed to emphasize the focus of the downstream processing to certain relevant regions in the images, thereby removing redundant information and noise in the workflow. Subsequently, task-specific pathological prior knowledge is leveraged to construct a meaningful entity-graph representation for the TRoIs. The entity-graphs are processed by the proposed hierarchical GNN to map the entity attributes and topology to the expected outcome of the pathology task. In this course, the neural network aims to learn meaningful representations for all the entities in the graph by incorporating their local neighborhood topology and neighboring entities' attributes. Following the learning between the tissue structure and functionality, the graph explainer is employed to perform post-hoc interpretation of the learning for each TRoI. Formally, the explainer aspires to identify influencing entities in the entity-graph that drive the mapping between the tissue structure and functionality. The explainer generates an explanation for each TRoI by highlighting the important elements in the entity-graph. As the explanations are demonstrated at a biologically defined entity-level, the pathologists can comprehend and evaluate the consistency between pathological understanding and the deep neural network's interpretation. Such bridging of the gap between AI and a pathologist can strengthen human trust in machines.

In this work, we employ the proposed HACT graph representation and HACT-Net, respectively, to hierarchically represent and learn tissue information in TRoIs. To elaborate and evaluate the methodology, we focus on the task of classifying Hematoxylin and Eosin (H&E)-stained breast cancer RoIs. Notably, our framework is generic enough to be applied to other cancer types (prostate, colon, etc.), staining techniques (HER2, ER, PR, multiplexed tissue images, etc.), tissue aspiration techniques (tissue microarray, biopsy, etc.), and image scales (tumor region, WSI, etc.).

2.1. *Semantic graph representation*

A tissue specimen consists of different types of entity sets available at different scales, such as cell, tissue region, and gland, that depict multivariate information about the tissue composition. For instance, cells as the entity set can provide information about cell morphology and the low-level cell microenvironment, whereas tissue regions as the entities can provide a high-level regional distribution of tissue and the tissue microarchitecture. Notably, the different types of entity sets cater both correlated and complementary information. Thus, the utilization of only a single-entity set to describe the tissue composition is incomplete, and AI algorithms designed using such a description are suboptimal. Alternatively, the utilization of multiple types of entities and their inter-dependence can capture more detailed semantics of the tissue structure, thereby supporting the efficacy of reliant AI solutions. The AI algorithm can operate on the tissue in a hierarchical order, e.g., fine to coarse scale, and leverage the information in one scale to localize the algorithm's attention onto relevant patterns in another scale. Such an approach resembles a pathologist's diagnostic procedure, where the pathologist begins the analysis of a tissue at a coarse scale, identifies relevant regions of interest, and pursues further analysis into the specific regions at a finer scale.

We adopt the aforementioned hypothesis to build a hierarchical graph representation, using multiple types of entity sets, to describe a tissue. Such graph representation considers the distribution of the entity types as a hierarchical organization, where lower scale entities formulate higher scale entities. The graph encodes per-entity local information in its nodes and embeds the entity-to-entity relationships in the edges. Each type of entity set constitutes a single-scale graph encoding morphological and intra-entity information. Further, the inter-entity interactions between two scales are captured via surjective hierarchies between the entity sets of the two scales. Specifically, our hierarchical entity-graph consists of three major components, (1) a *CG* to characterize low-level cell information, where the nodes and edges of the CG depict the cells and cellular interactions, respectively, (2) a high-level *Tissue-graph* (TG), where the nodes and edges of the TG depict the tissue regions and their interactions, respectively, and (3) a set of cell-to-tissue hierarchies, utilizing the spatial distribution of the cells with respect to tissue regions. The combination of the three components provides a *HACT* representation [64] as presented in Figure 2. The visual attributes of cells and tissue regions are encoded in the nodes of CG and TG, respectively. HACT, by utilizing a multiset of entities, captures both the cellular and tissue microenvironments and includes both local and global tissue composition information. Compared to the pixel-based representation, i.e., an image, of a tissue specimen, the

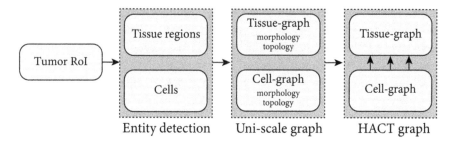

Figure 2:   Overview of the proposed HACT entity-graph representation of an input tumor region-of-interest.

HACT representation is compact and memory-efficient, and it can thus describe arbitrarily large tissue specimens. We select the set of cells and tissue regions considering their significance in breast tumor subtyping. For other tasks and tissue types, different relevant multisets of entities can be selected. For instance, to assess the prognostic value of tumor-infiltrating lymphocytes in triple-negative breast cancer patients, the CG in HACT may consist only of tumor infiltrating lymphocytes and cancer cells to describe the cancer cell microenvironment, and the TG in HACT may then consist of epithelium to represent the tumor microenvironment. Details of CG, TG, cell-to-tissue hierarchies, and HACT representation construction are presented in the following subsections.

2.1.1. *CG representation*

A *CG* characterizes low-level cell information, where the nodes represent cells and edges encode cellular interactions. In this chapter, we demonstrate the HACT graph representation on H&E-stained breast TRoIs. In H&E staining, the hematoxylin stains cells nuclei blue, and eosin stains the extracellular matrix and cytoplasm pink, with other structures taking on different shades, hues, and combinations of these colors. Thus, the nodes and edges in our CG constitute cell nuclei and nuclear interactions, respectively. To construct a CG for a TRoI, first, the nodes are identified, i.e., nuclei, using any standard nuclei detection network. In this work, we employ the Hover-Net model [22], a state-of-the-art nuclei segmentation network, pre-trained on a multi-organ nuclei segmentation dataset [21], for the nuclei detection. HoVer-Net leverages the instance-rich

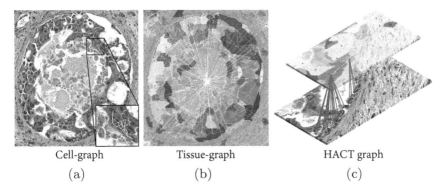

Cell-graph	Tissue-graph	HACT graph
(a)	(b)	(c)

Figure 3:   Visualizing (a) CG, (b) TG, and (c) HACT entity-graph representations. Nodes are presented in red and edges in yellow. Cell-to-tissue hierarchies are shown in blue in HACT. Note that not all hierarchies in the HACT graph are shown, for sake of visual clarity.

information encoded within the vertical and horizontal distances of nuclear pixels to their centers of mass. These distances are utilized to separate clustered nuclei, resulting in an accurate segmentation, particularly in areas with overlapping instances. Following nuclei detection, we extract hand-crafted features or DL features for the detected nuclei to encode their visual attributes in the CG nodes.

The CG topology can be heuristically defined to represent cellular interactions. We hypothesize that spatially near-by nuclei interact and should be connected in CG, and distant nuclei with much weaker interactions may remain disconnected in the CG. To this end, we use the k-Nearest Neighbors (k-NN) algorithm to build the initial topology and prune the k-NN graph by removing edges longer than a heuristically defined distance threshold. We use Euclidean distance between the nuclei centroids in the image space to quantify the nuclear distance. Figure 3(a) displays a sample CG elucidating the nodes and edges in the zoomed-in sub-image.

2.1.2. *TG representation*

A TG depicts a high-level tissue microenvironment, where the nodes and edges of TG represent the tissue regions and their interactions, respectively. Similar to CG, TG are constructed by first identifying tissue components, such as epithelium, stroma, necrosis, adipose tissue, and lumen, followed by feature representation of the tissue regions and TG topology construction. In this chapter, we identify the tissue regions in an unsupervised approach. First, the TRoI is oversegmented at lower magnification to detect non-overlapping homogeneous regions, i.e., superpixels. The superpixels are the result of perceptual grouping of pixels and carry more information than image pixels. To this end, we employ the simple linear iterative clustering (SLIC) algorithm [84] by emphasizing space proximity. Subsequently, to minimize redundancy in superpixels and to achieve meaningfully compact tissue components, we hierarchically merge neighboring superpixels containing similar color attributes. The merged superpixels denote the tissue regions and correspondingly depict the nodes in TG. To represent the visual attributes of the detected tissue regions, we extract hand-crafted or DL features at high magnification. Additionally, spatial centroids of the tissue regions normalized by the image size are included to construct feature representations of the tissue regions. To generate the TG topology, we assume that adjacent tissue components biologically interact and should be connected. To this end, we construct a region adjacency graph (RAG) [85] using the spatial centroids of the superpixels. The feature representations define the initial node features, and the RAG edges define the TG edges to accommodate the tissue microenvironment. Figure 3(b) presents a sample TG, where the large node, at the center, denotes the centroid of the surrounding stroma that is connected to epithelium and necrotic regions.

2.1.3. *HACT representation*

To jointly represent the low-level CG and high-level TG for a tissue, we introduce the HACT graph representation, defined as a topological mapping between the CG and TG nodes. We implement the mapping by using the spatial distribution of the nuclei with respect to the tissue regions, i.e., the mapping is positive if the nucleus represented by a specific node in the CG spatially belongs to the tissue region represented by the corresponding node in the TG. An overview of HACT in Figure 3(c) displays a sample multi-scale graph and the corresponding hierarchies. Notably, the hierarchical representation enables the accessibility of the inter-level topological relation during the mapping of the tissue structure to tissue function.

2.2. *HACT-Net*

The HACT graph representation is processed by a hierarchical GNN, HACT-Net to map the tissue structure to function relationship. In this chapter, HACT-Net aims to map the HACT representation of a TRoI to the corresponding breast cancer subtype. The overview of HACT-Net workflow is demonstrated in Figure 4. Formally, HACT-Net consists of two GNNs, i.e., a *Cell-GNN* (CG-GNN) and a *Tissue-GNN* (TG-GNN), to hierarchically process a HACT graph from fine-to-coarse scale. The proposed HACT-Net can work with any GNN architecture [86–93]. Specifically, we use the Graph Isomorphism Network (GIN) [91], an instance of a message passing neural network [86] that allows a fixed-size discriminative representation to be learnt for an input graph. We employ the GIN layers in both CG-GNN and TG-GNN architectures. First, we apply CG-GNN to the CG in an HACT representation to learn contextualized cell-node representations. For each node in a CG, we update the feature representation by considering the set of neighboring nodes' features in an iterative fashion through the GIN layers in CG-GNN. The output feature representations of the cell-nodes are concatenated with the initial TG node representations by utilizing the cell-to-tissue hierarchies to initialize the TG input to TG-GNN. Subsequently, TG-GNN processes the TG to output contextualized tissue-node representations. Notably, the final tissue-node representations encode both local nuclei and global tissue region information. To adapt to different sub-graph structures in CG and TG, we employ a jumping-knowledge technique [94] to the node features in CG-GNN and TG-GNN. TG-GNN outputs the contextualized fixed-size graph-level features, which are

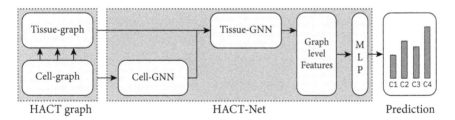

Figure 4:   Overview of HACT-Net workflow. A HACT graph representation is processed by a hierarchical GNN, HACT-Net, to predict the cancer subtype of the input tumor region-of-interest.

subsequently processed by a multi-layer perceptron (MLP) classifier to output breast cancer subtypes.

2.3. Explaining the network prediction

The HACT entity-graph representation offers a novel way of encoding a tissue composition using pathologically comprehensible tissue constituents. While being a strong argument to enforce trust between pathologists and AI agents, the HACT-Net, responsible for mapping an HACT representation to a cancer subtype, remains a deep neural network without any direct way of understanding its decision-making procedure. For this reason, we often refer to deep neural networks as black-box models. Therefore, explainable DL technologies in computational pathology are of paramount interest to promote the employment of AI in clinical settings [74]. In the context of computational pathology, explainability is defined as making the DL decisions understandable to pathologists [75]. Figure 5 presents the proposed explainability workflow: a graph explainer uses the prediction from the HistoCartography module to highlight regions in the input TRoI, in terms of entity-wise importance, which

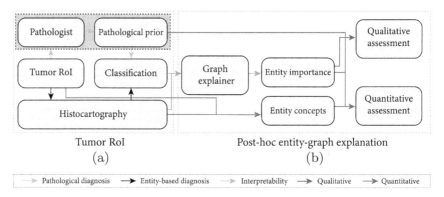

Figure 5: Overview of the proposed concept-based explainability framework. (a) presents the processing of a histology image by a pathologist and HistoCartography, the proposed entity-graph approach. (b) presents a post-hoc graph explainability technique, producing an entity-level importance score, to generate a qualitative explanation of the AI solution. On utilizing the entity-level comprehensible pathological concepts, prior pathological knowledge, and entity-level importance scores, our framework can produce quantifiable explanations in pathologically understandable terminologies.

drives the network prediction. The graph explainer produces a per-instance explanation (in terms of entity importance) that can be used for qualitative assessment. This can also be used for quantitative evaluation when coupled with pathological priors and the extracted entity concepts.

2.3.1. *Feature attribution explainers*

Recently, several research works have been devoted to demystifying the concept representations of CNNs in automated diagnosis. Patch-level explainability methods [76, 77] build patch-level *heatmaps,* where an importance score is computed per pixel to identify the regions of importance. For instance, [77] use layerwise relevance propagation [78] to generate positive scores for pixels that are positively correlated with the class label and negative scores otherwise. Such approaches have several limitations. First, pixel-level heatmaps fail to capture the spatial organization and interactions of relevant biological entities. Second, pixel-level analysis is completely detached from pathological guidelines recommended for decision-making. Third, pixel-level explainability results in blurry heatmaps, which inhibits the identification of the relevant set of tissue constituents and their interactions. Figure 6 (left) highlights the limitations of pixel-level heatmap explanation.

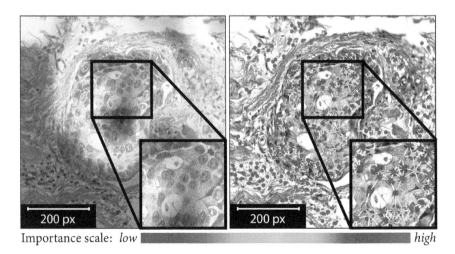

Figure 6: Sample explanations produced by pixel- and entity-based explainability techniques for a DCIS region of interest.

Following the discussion in previous sections, cancer diagnosis, prognosis, and therapeutic response prediction using a histology image are all heavily influenced by the underlying tissue composition. To that end, we have pursued an entity-graph-based methodology to semantically represent and learn the tissue composition of a histology image, utilizing relevant information from the involved semantic entities and their interactions. A graph-learning mechanism, e.g., a GNN, learns and maps the tissue semantics to tissue functionality, e.g., breast tumor subtype in this chapter. Such a paradigm shift from pixel- to entity-wise processing provides more control over the analysis of entity-level pathological concepts, such as shape, size, and chromaticity. An entity-graph representation consisting of biological entities, already provides an *interpretable* input space to pathologists, that they can relate to and reason with. Thus, an entity-level explainability method can highlight any relevant subset of entities in an entity-graph, the entity-level concepts of which can then be assessed by a pathologist to decipher any black-box deep neural network's decision-making procedure. If the generated entity-level explanation, i.e., the subset of entities identified by the explainability method, resonate with the pathologically relevant set of entities for the TRoI, then the trust between the pathologists and the entity-based AI solutions can be enhanced. Further, the flexibility of incorporating a task-specific multiset of entities into the entity-graph generation enables the pathologists to evaluate the consistency of an AI solution's decision-making procedure with the prior pathological knowledge.

In this chapter, for simplicity, we constrain our explainability analysis to single-scale CG representations for breast TRoIs. Note that a similar approach can be extended to other entity sets, e.g., TG, gland-graph, multi-scale entity-graph, etc. Also, we focus on *post-hoc feature-attribution*-based graph explainability techniques [79–82] among several explainability methods for graph-structured data. The studied graph explainer can (i) explain any prediction made with a pre-trained GNN, and (ii) assign an importance score to each node in the input entity-graph. For instance, in Figure 6 (right), we show a CG-based heatmap that directly provides the importance of each nuclei. Important nuclei (in red) consist of cancerous nuclei located inside the gland, whereas the low-importance nuclei set (blue) consists of normal epithelial nuclei, stromal nuclei, and lymphocytes that span the periphery and outside the glandular region. In this work, we study three classes of feature attribution methods, namely: gradient-based (class activation maps (CAM), specifically,

GraphGrad-CAM, GraphGrad-CAM++ [81]); layerwise-relevance propagation (*GraphLRP* [82]); and graph pruning-based (*GNNExplainer* [79,
80]), *all three are briefly* introduced in what follows. The reader can refer
to [72] for a formal and detailed description.

- **Gradient-based approaches [81]** utilize the gradient strength of
  the predicted tumor type with respect to deep layers as a means to
  quantify the importance of each deep feature. Deep feature importance scores can then be easily mapped back to the input to derive
  input-level, i.e., nuclei-level, importance scores. An extension of
  GraphGrad-CAM can be made to improve the spatial localization of
  the deep feature importance scores. This method is denoted as
  GraphGrad-CAM++.
- **Layerwise relevance propagation (LRP) [82]** is based on the propagation of the output prediction confidence score, i.e., the logit of the
  predicted tumor type, backward in the DL network. Propagation rules
  are used to quantify the positive contribution of each input element,
  i.e., of each nucleus, toward a certain prediction. This approach is
  referred to as *GraphLRP*.
- **Graph pruning** is a model-agnostic explainability technique for
  GNNs, referred to as *GNNExplainer* ([79, 80]). *GNNExplainer* aims
  to find the smallest sub-graph, i.e., the explanation, such that the network prediction on this sub-graph remains the same as the original
  graph. The optimal sub-graph is found using an optimization problem
  that learns a mask over the nodes that will either activate or deactivate
  them. The *GNNExplainer* can be seen as a feature attribution technique with binarized node importance scores.

Feature attribution methods on graph-structured data remain an open
and active research direction. In other words, there is no unique and
established approach to assign feature-level importance scores, e.g.,
nuclei-level scores in the context of CG explanations. For each entity-
graph, we can obtain different explanations by employing different feature attribution techniques. Consequently, the outcomes of feature
attribution techniques still need to be analyzed, understood, and validated by expert pathologists. Thus, identifying the explainability method
that provides the most suitable explanations for several RoIs becomes
crucial. In Figure 7, we demonstrate qualitative explanations, i.e., nuclei-
level importance scores, from the four studied graph explainers. We

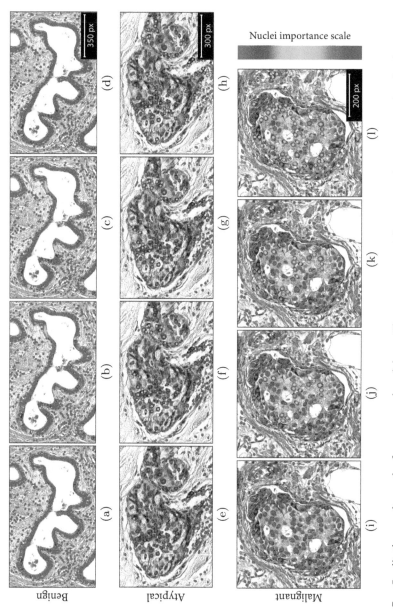

Figure 7:   Qualitative results on the four proposed explainers. The rows represent the cancer subtypes, grouped under three categories in this example as: Benign, Atypical, and Malignant. The columns represent the graph explainability techniques: *GNNExplainer; GraphGrad-CAM, GraphGrad-CAM++,* and *GraphLRP*. Nuclei-level importance ranges from blue (the least important) to red (the most important).

observe that *Graph Grad-CAM* and *GraphGrad-CAM++* produce similar importance maps. The highlighted regions are spatially defined and correspond to areas with similar nucleus types. The *GNNExplainer* generates almost binarized nuclei importance scores. Interestingly, even though the gradient-based and pruning-based techniques operate differently, they seem to consistently highlight similar regions. Indeed, both approaches focus on the relevant epithelial region and defocus on stromal nuclei and lymphocytes outside the glands. In contrast, *GraphLRP* produces less interpretable maps through either high spatial localization or low spatial localization. A qualitative visual assessment suggests that (i) not all explainers are equal, while some explainers consistently highlight the same regions, e.g., *GNNExplainer* and *GraphGrad-CAM*, others like the *GraphLRP* provide different explanations, and (ii) feature attribution methods should be further analyzed to identify better explainers.

2.3.2. *Entity-level concept analysis*

A feature attribution-based graph explainer outputs entity-level importance scores for an input entity-graph. To assess the quality of a generated explanation, qualitative evaluation of the explanation by pathologists is the candid measure. However, it requires evaluation by task-specific expert pathologists, which is subjective, time-consuming, cumbersome, and expensive. Further, the generated importance scores are typically incomprehensible to a pathologist as such scores do not express any explanation in pathologically understandable terminologies, e.g., the density, shape, or size of the cells, which the pathologists typically refer to in their own training, assessments, and discussions. Such disparity complicates a comprehensive analysis. Additionally, different explainers adopt different mechanisms for scoring entity importance, which further limits any direct comparison of explanations from different explainers.

To address the aforementioned challenges, we propose a quantitative framework [72], given in Figure 8, to objectively quantify the quality of a graph explainer and express the explanations in pathologically understandable terminology. The quantification utilizes the set of TRoIs, explainer generated entity-level importance scores, pathologically relevant entity-level concepts, and task-specific prior pathological knowledge over entity-level concepts. The task-specific entity-level concepts, defined by pathologists, describe the morphology and local topology of

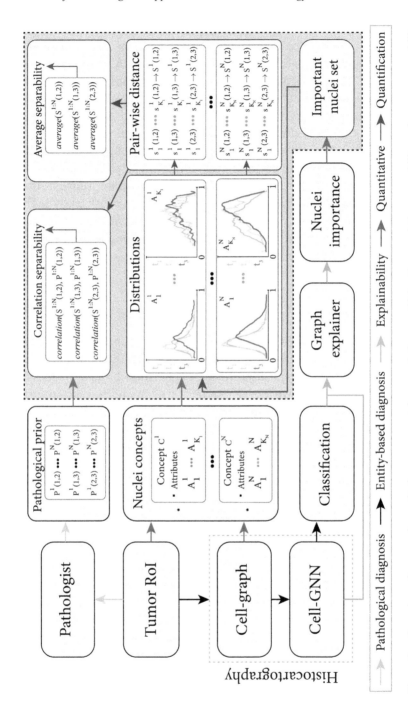

Figure 8:   Overview of proposed quantitative assessment framework for quantifying the correlation separability and average separability metrics for CG explanations.

the entities. More specifically, the operation of the quantitative framework can be described as follows:

(1) For each TRoI, the framework identifies an important set of entities by thresholding the entity-level importance scores. The entity-level concept values are computed for these important entities.

(2) For each class, the important set of entities from all the tumor RoIs belonging to the given class are collected to describe the global set of relevant entities corresponding to the class.

(3) For each class, concept-wise distributions are created using the corresponding global set of relevant entities.

(4) For each concept and each unique pair of classes, a distance metric between their distributions is computed. Such distance quantifies the separability between the pair of classes while using the given concept. The concept-wise class separability scores can be used to compare the relevance of the individual concept, with larger class separability scores indicating more relevant concepts.

(5) For a pair of classes, we define a *correlation separability* metric by computing the Pearson correlation between the concept-wise class-separability scores and the concept-wise pathological prior. The pathological prior signifies the relevance of each concept for differentiating the considered pair of classes. Such prior can be defined by an individual pathologist or by the consensus of several pathologists. Notably, the prior is independent of the graph explainer algorithm and the quantitative framework. The correlation is high if the identified set of important entities and the respective concepts are consistent with the pathological understanding of the considered classification task. Inherently, the agreement conveys and quantifies the relevance of the graph explainer that generates the entity-level importance scores.

(6) For a pair of classes, we define an *average separability* metric by computing the average of the concept-wise class-separability scores. This metric signifies the expected separability between a pair of classes for a particular graph explainer.

We herein demonstrate the aforementioned explanation quantification framework and procedure based on a CG-based breast cancer subtyping task, as illustrated in Figure 8. Table 1 lists the pathologically relevant set of nuclei-level concepts considered herein, together

Table 1:   List of (*column 1*) pathologically relevant nuclear concepts considered for breast cancer subtyping and (*column 2*) the measured attributes for each nuclear concept. The expected behavior for each concept is presented in (*columns 3, 4, 5*) to categorize a tumor into three breast cancer subtypes, Benign, Atypical, and Malignant.

Concept	Attribute	Benign	Atypical	Malignant
Size	Area	Small	Small-medium	Medium-large
Shape	Perimeter	Smooth	Mild irregular	Irregular
	Roughness			
	Eccentricity			
	Circularity			
Shape variation	Shape factor	Monomorphic	Monomorphic	Pleomorphic
Spacing	Mean spacing	Evenly crowded	Evenly spaced	Variable
	Standard deviation spacing			
Chromatin	GLCM dissimilarity	Light euchromatic	Hyperchromatic	Vesicular
	GLCM contrast			
	GLCM homogeneity			
	GLCM ASM			
	GLCM entropy			
	GLCM variance			

with their expected behavior for each cancer subtype. For instance, the nuclei important in an invasive TRoI, i.e., cancerous nuclei, are expected to be of a larger size than the normal epithelial nuclei in a benign TRoI. Following Step (1) above, we normalize the entity-level importance scores, and select the important set of nuclei by thresholding with a pre-defined threshold value. The comprehensible concepts are estimated using the concept-wise set of attributes, given in Table 1. For example, to characterize the nuclei shape, we measure its perimeter, roughness, eccentricity, and circularity. Notably, to support inter-attribute comparisons, we conduct attribute-wise normalization across all entities from all TRoIs in the dataset. Following Step (2), we build the global set of important nuclei for each cancer subtype by collecting the important sets of nuclei from all the TRoIs belonging to each cancer subtype. In Step (3), the global set of nuclei attributes for each cancer subtype are used to construct histograms per attribute and per cancer subtype. In Step (4), for each attribute and all unique pairs of cancer subtypes, we define class separability as the optimal transport between their cumulative histograms; accordingly, we compute the Wasserstein distance between histograms as a class separability score. The measured attribute-wise class separability scores corresponding to a concept for a pair of classes are averaged to describe the class separability between the two classes while using that particular concept. Following Steps (5) and (6), we compute *correlation separability* and *average separability* metrics for each unique pair of cancer subtypes. The pathological prior is obtained by the consensus of three pathologists, from different university hospitals. They were asked to rank each of the nuclear concepts in terms of their relevance toward discriminating each pair of cancer subtypes. In this study, we focus on a three-class scenario (Benign, Atypical, Malignant), leading to three unique pairs of classes, i.e., Benign vs. Atypical, Benign vs. Malignant, and Atypical vs. Malignant. For instance, for Benign vs. Atypical, given an Atypical TRoI we asked the pathologists to rank the relevance of the"nuclear shape" concept, among the set of pre-defined concepts, for annotating the TRoI as Atypical instead of Benign. The computed pair-wise *correlation separability* metrics and the *average separability* metrics are aggregated to define the overall *correlation separability* metric and the overall *average separability* metric, respectively.

3. Experiments and Results

We experiment and evaluate the proposed classification performance of the HistoCartography methodology on the BReAst Carcinoma Subtyping (BRACS) dataset, an inhouse large cohort of breast TRoIs extracted from H&E-stained breast carcinoma WSIs. We compare the classification performance of the proposed methodology with recent CNN and GNN approaches for breast cancer multi-class subtyping. We further benchmark the classification performance against the independent re-evaluation of the dataset by three expert pathologists. The explanation quality is further evaluated by additional external pathologists, in order to study the agreement between the proposed approach and the experts.

3.1. *BReAst Carcinoma Subtyping (BRACS) dataset*

The BRACS dataset consists of 4391 TRoIs acquired from 325 H&E-stained breast carcinoma WSIs. BRACS is publicly available[2] and was collaboratively created by the National Cancer Institute-IRCCS-Fondazione Pascale, Naples, Italy; the Institute for High Performance Computing and Networking-CNR, Naples, Italy; and IBM Research Europe. Specifically, the WSIs were selected from the archives of the Department of Pathology at the National Cancer Institute — IRCCS Fondazione Pascale. The WSIs were scanned with an Aperio AT2 scanner at 0.25 μm/pixel for 40x resolution. The selected TRoIs were annotated as Normal, Benign, Usual ductal hyperplasia (UDH), Atypical ductal hyperplasia (ADH), Flat epithelial atypia (FEA), DCIS, and Invasive. The TRoIs were selected such that all the classes were equally represented. In the scenario used to study the explainers, the problem is simplified to a three-class scenario as Benign (B) that includes Normal, Benign, and UDH TRoIs; Atypical (A) with ADH and FEA RoIs; and Malignant (M) with DCIS and Invasive TRoIs. Each TRoI was annotated independently by three pathologists. TRoIs with disagreements were further discussed and annotated by the consensus of the three pathologists. Note that the pathologists had access to the entire WSI while annotating the TRoIs, thus utilizing WSI-level context information during the annotation procedure. The BRACS dataset was built to highlight real-life tumor variability that pathologists encounter in their daily practice. Thus, the dataset includes

---

[2]BRACS: BReAst Carcinoma Subtyping: https://www.bracs.icar.cnr.it/.

TRoIs possessing variable dimensions, staining and appearance variability, tissue preparation artifacts, multiple patterns from each cancer subtype, etc. The artifacts include tissue-folds, tears, ink stains, blur, and others. To exemplify the per cancer subtype pattern variability, the DCIS subtype includes variable grades, i.e., low-grade, moderate-grade, and high-grade DCIS, thus including papillary, cribriform, solid, and comedo architectures. For developing AI solutions to classify the breast TRoIs in the BRACS dataset, we split the dataset into train, validation, and test sets. The TRoIs were split at the WSI-level, so that no two TRoIs from the same WSI belong to different splits. The AI solutions use train and validation splits during the training phase and evaluate the trained solution's generalizability on the independent test split. We provide the train, validation, and test splits of our presented results in [98] for reproducibility and a fair comparison with future AI solutions.

3.2. *Pre-processing*

The H&E-stained specimen exhibit appearance variability for various reasons, such as specimen preparation method, staining protocols (e.g., temperature of the adopted solutions), fixation characteristics, and imaging device characteristics. While being part of the challenges that an AI solution needs to address, such variability in appearance may adversely impact the model performance. Thus, we begin our methodology by reducing the appearance variability in the RoIs. To this end, we employ the unsupervised stain normalization algorithm proposed in [83]. The algorithm does not involve training of model parameters and is computationally inexpensive. We further improve workflow integration by employing a scalable and fast pipeline to implement the above algorithm as proposed in [42].

3.3. *Breast cancer subtyping results*

The proposed HistoCartography methodology is employed to classify the TRoIs into respective breast cancer subtypes. First, HACT representations are extracted, and then, HACT-Net is utilized for the classification task. HACT-Net is trained by minimizing the mean cross-entropy loss between the output logits and the ground truth cancer subtypes for all the TRoIs in the BRACS training set. Considering the per-class data imbalance,

270 Artificial Intelligence Applications in Human Pathology

weighted F1-score is used to quantify the classification performance on the test set. The details of the network architecture, the training hyperparameters, and the ablation experiments are described in [64]. We compare the HACT-Net's classification performance with a series of DL-based baselines including CNN and GNN-based approaches. The mechanism of each baseline is presented briefly as follows:

- *Single-Scale CNN*: This baseline uses a CNN to process TRoIs in a patch-wise manner [53]. The patches from TRoIs are extracted at a single magnification, i.e., 10x, 20x, 40x. An empirical evaluation concluded that the best classification performance is achieved for processing the patches extracted at 10x magnification.
- *Multi-Scale CNN*: This baseline uses a multi-scale CNN to process the TRoIs in a patch-wise manner [53]. The approach extracts concentric patches of fixed-size from multiple magnifications. A shared CNN processes a stream of concentric patches altogether, to acquire tissue context information from multiple magnifications. The multi-scale output representations from CNNs are merged using the "late fusion with single stream + LSTM" strategy [53], and the merged representation is used to classify the patches. Through ablation experiments, we extract patches of size 128×128 pixels from 10x, 20x, and 40x magnifications to train this multi-scale CNN.
- *Cell-GNN*: This baseline uses only the CG representations to classify the TRoIs. The objective of the baseline is to compare the proposed hierarchical representation and learning with standalone CG-based learning.
- *Tissue-GNN*: This baseline uses only the TG representations to classify the TRoIs. The objective of the baseline is to compare the proposed hierarchical representation and learning with standalone TG-based learning.

We also benchmark the HACT-Net classification performance against the pathologists' performance on the BRACS test set. We followed the evaluation protocols in [5], to assess the performance of three board-certified pathologists (excluding the pathologists providing the original annotations). The participating pathologists are from three medical centers, (1) National Cancer Institute — IRCCS Fondazione Pascale, Naples, Italy, (2) Lausanne University Hospital, CHUV, Lausanne, Switzerland,

and (3) Aurigen — Centre de Pathologie, Lausanne, Switzerland. They are specialized in breast pathology and have been in practice for over 20 years. The pathologists independently and remotely annotated the TRoIs in the BRACS test set, without having access to the corresponding WSIs. Such evaluation protocol ensures equal field-of-view for all the pathologists as well as the DL-based methods. The per-class performance and aggregated classification performance for all the baselines, HACT-Net, and the pathologists are presented in Table 2.

We observe that the Single-Scale CNN produces the lowest aggregated weighted-F1 score, followed by the Multi-Scale CNN. All the graph-based baselines outperform the CNN baselines. HACT-Net significantly outperforms the CNN and GNN baselines. The complementary multi-scale information, from low-level CG and high-level TG, is effectively utilized by HACT-Net to improve the class-wise and overall classification performance. The class-wise performance analysis in Table 2 shows that the Invasive category is the easiest to detect. It translates to topologically recognizable patterns with scattered nodes and edges in the CG and TG. UDH and ADH TRoIs are the most difficult to model, partially as they have a high intra-class variability and high inter-class ambiguity with Benign and DCIS TRoIs. Standalone cell-level and tissue-level information are not sufficient to categorize both UDH and ADH RoIs. This is evident from the complementary performance of Cell-GNN and Tissue-GNN in Table 2. Cell-GNN uses the monomorphic cell information and the diameter of CG as a measure of gland size, to better encode the ADH samples. In contrast, the Tissue-GNN uses the information of apical snouts from TGs to identify FEA TRoIs. HACT-Net having access to both cell and tissue microenvironment information better encodes the ADH and FEA RoIs, thereby resulting in better performance. Similar behavior is observed in HACT-Net performance for other cancer subtypes.

Table 2 indicates that HACT-Net provides comparable performance with respect to the pathologists for Benign, ADH, DCIS, and Invasive subtypes, and results in significantly better performance for the Normal and FEA categories. Overall, HACT-Net outperforms the pathologists in identifying pre-cancerous atypical subtypes, which is a crucial and challenging task, while managing equivalent performance on the relatively easier benign and cancerous subtypes. Table 3 presents the inter-observer concordance rates for the BRACS test set. We notice significant difference in concordance rates between pathologist (1, 2) and pathologist

Table 2:   Mean and standard deviation of class-wise F1-scores, and 7-class weighted F1-scores. Results expressed in %. For the methods, the standard deviations represent variation across training runs with different random initializations; and for the pathologists, the inter-observer variability.

	Normal	Benign	UDH	ADH	FEA	DCIS	Invasive	Weighted-F1
Single-Scale CNN	48.67 ± 1.71	44.33 ± 1.89	45.00 ± 4.97	24.00 ± 2.83	47.00 ± 4.32	53.33 ± 2.62	86.67 ± 2.64	50.85 ± 2.64
Multi-Scale CNN	50.33 ± 0.94	44.33 ± 1.25	41.33 ± 2.49	31.67 ± 3.30	51.67 ± 3.09	57.33 ± 0.94	86.00 ± 1.41	52.83 ± 1.92
Cell-GNN	58.77 ± 6.82	40.87 ± 3.05	46.82 ± 1.95	39.99 ± 3.56	63.75 ± 10.48	53.81 ± 3.89	81.06 ± 3.33	55.94 ± 1.01
Tissue-GNN	63.59 ± 4.88	47.73 ± 2.87	39.41 ± 4.70	28.51 ± 4.29	72.15 ± 1.35	54.57 ± 2.23	82.21 ± 3.99	56.62 ± 1.35
HACT-Net	61.56 ± 2.15	47.49 ± 2.94	43.60 ± 1.86	40.42 ± 2.55	74.22 ± 1.41	66.44 ± 2.57	88.40 ± 0.19	61.53 ± 0.87
Pathologist	51.57 ± 11.70	52.15 ± 1.85	38.78 ± 10.22	37.89 ± 3.12	46.16 ± 19.64	68.26 ± 2.62	92.49 ± 2.14	56.36 ± 0.76

Table 3: Concordance among independent pathologists' annotations. Results expressed in %.

	Pathologist 1	Pathologist 2	Pathologist 3	Ground truth
Pathologist 1	—	47.60	50.96	56.71
Pathologist 2	—	—	64.38	57.99
Pathologist 3	—	—	—	56.55

(2, 3), and comparable concordance between pathologist (1, 2) and pathologist (1, 3). The observation can be reasoned to diagnostic protocol differences across different regions, i.e., Pathologist 1 from Naples, Italy, vs. 2 and 3 from Lausanne, Switzerland.

3.4. *Breast cancer explainability results*

To test the proposed novel domain-concept-based quantitative metrics, we analyze two key components: the histogram distributions and the class separability scores. Figure 9 illustrates the per-class histograms for each explainer and the best attribute per concept. The best attribute for a concept is the one with the highest pair-wise class separability. The row-wise observation illustrates that the *GNNExplainer* and the *GraphLRP* provide, respectively, the maximum and the minimum pair-wise class separation. The histograms for a concept and for an explainer can be analyzed to assess the agreement between the selected nuclei concept for importance, and the expected concept behavior. This is presented in Table 1, for all the classes. For instance, nuclear area is expected to be higher for malignant TRoIs than for benign ones. The area histograms for *GNNExplainer*, *GraphGrad-CAM*, and *GraphGrad-CAM++* indicate that the nuclei set important for malignant TRoIs includes nuclei with higher area compared to that for benign TRoIs. Similarly, the important nuclei in malignant TRoIs are expected to be vesicular, i.e., exhibit high texture entropy, as opposed to being light euchromatic, i.e., showing moderate texture entropy, in benign TRoIs. The chromaticity histograms for *GNNExplainer*, *GraphGrad-CAM*, and *GraphGrad-CAM++* display this behavior. Additionally, the histogram analysis can help reveal any important concepts and attributes. For instance, nuclear density proves to be the least important concept for differentiating the considered classes.

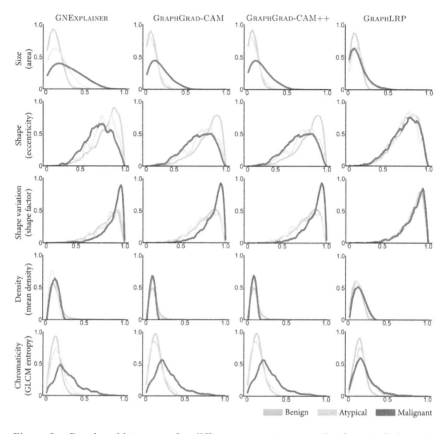

Figure 9:   Per-class histograms for different *concepts* across the four studied graph explainers: *GNNExplainer*, *GraphGrad-CAM*, *GraphGrad-CAM++*, and *GraphLRP*.

Table 4 provides the per-class pair and aggregate concept-based average and correlation separability scores. We observe that, first, the *GNNExplainer* has the best average separability score followed by *GraphGrad-CAM++* and *GraphGrad-CAM*. All explainers outperform a baseline (*Random*) with random nuclei selection, which conveys that the quality of the explainers' explanations is better than random. We note that *GraphGrad-CAM* and *GraphGrad-CAM++* quantitatively perform very similarly, which is consistent with our qualitative analysis in Figure 7. Looking at the correlation separability scores, we observe that GNNExplainer, *GraphGrad-CAM*, and *GraphGrad-CAM++* have a positive correlation separability score for classifying the class pairs Benign vs.

Table 4:   Per class pair and aggregate concept-based average and correlation separability scores. B stands for Benign, A for Atypical, and M for Malignant. The first- and second-best values are indicated in **bold** and <u>underlined</u>, respectively.

Metric	Explainer	B vs. A (↑)	B vs. M (↑)	A vs. M (↑)	Aggregate (↑)
Average separability	GNNExplainer	**1.54**	**2.78**	1.93	**6.25**
	GraphGrad-CAM	1.15	2.57	<u>2.08</u>	5.80
	GraphGrad-CAM++	1.18	<u>2.58</u>	**2.09**	<u>5.85</u>
	GraphLRP	<u>1.38</u>	1.59	1.47	4.44
	Random	1.05	1.00	0.95	3.00
Correlation separability	GNNExplainer	−0.02	0.36	0.38	0.72
	GraphGrad-CAM	<u>−0.01</u>	<u>0.57</u>	<u>0.58</u>	<u>1.14</u>
	GraphGrad-CAM++	**−0.01**	**0.58**	**0.59**	**1.16**
	GraphLRP	−0.15	−0.49	−0.23	−0.87
	Random	−0.37	−0.31	−0.18	−0.86

Malignant and Atypical vs. Malignant, and nearly zero values for the class pair Benign vs. Atypical. This observation is an indicator that the explanations for Benign vs. Malignant and Atypical vs. Malignant bear similar relevance to concepts as the pathologists, while a different concept relevance is offered for the class pair Benign vs. Atypical. *GraphGrad-CAM++* has the best overall agreement at the concept-level with the pathologists, followed by *GraphGrad-CAM* and *GNNExplainer*. The random baseline (*Random*) follows the three proposed explainers, while *GraphLRP* has the least agreement.

Table 5 highlights per concept separability scores for the best explainer *GNNExplainer*. The nuclear size is the most relevant concept, i.e., the concept with the highest separability score, followed by the nuclear chromaticity and the shape variation. Comparatively, the nuclear density is the least relevant concept as confirmed by Figure 8, where the class-wise histograms almost follow the same distribution.

4. Summary

AI holds the promise of strengthening the complex and demanding tasks of pathologists and has the potential to improve patient care. The current gap between the developed AI solutions and the diagnostic expectations

Table 5:   Quantification of concepts for pair-wise and aggregated class separability in the best explainer proposed: *GNNExplainer*. B stands for Benign, A for Atypical, and M for Malignant. The first- and second-best values are indicated in **bold** and <u>underlined</u>, respectively.

	B vs. A ($\uparrow$)	B vs. M ($\uparrow$)	A vs. M ($\uparrow$)	Aggregate ($\uparrow$)
Size	**3.26**	**6.24**	**3.47**	**12.97**
Shape	1.27	2.23	1.60	5.10
Shape variation	0.69	2.30	1.99	4.97
Density	1.01	0.80	0.52	2.33
Chromaticity	<u>1.44</u>	<u>2.31</u>	<u>2.07</u>	<u>5.82</u>

in clinical practice impedes the progress in adopting AI-based decision support tools by pathologists. In this chapter, we focus on two main aspects of such a challenge, namely building pathology-motivated AI solutions for analyzing histology images and the explainability of the AI's diagnostic procedure to pathologists. We argue that developing AI technologies that both leverage the power of data-driven ML techniques and align with the reasoning mechanisms in clinical practice holds the key to overcoming the adoption barrier and facilitating the integration of AI technology in the daily workflow of pathologists.

We propose the HistoCartography methodology to represent and analyze histology images in an entity-based manner to bridge the gap between AI and pathologists. The entity-based representation of a histology image captures domain-specific multi-scale semantic entities, the inter-scale entity-to-entity interactions, and intra-scale hierarchies between the entities. Further, the multi-scale topological tissue representation is processed by a proposed hierarchical graph-based AI technique, to map the tissue representation to the respective tissue functionality and phenotype. We evaluate the proposed methodology by subtyping breast cancer tumor regions-of-interest by incorporating cells and tissue regions as the biologically relevant entities. Accordingly, we term such entity-graph representation as a HACT graph and the graph-based AI technique as HACT-Net. Upon evaluation and comparison with several state-of-the-art AI-based approaches, we establish the classification efficacy of the proposed methodology. The HistoCartography methodology can be adapted to other pathological tasks across various tissue types. It can seamlessly scale to large regions-of-interest to encompass both local and global contextual tissue information for better classification. Additionally, it allows various task-specific entities and prior pathological knowledge to be incorporated in constructing task-specific comprehensive tissue representations.

To further understand the AI solution in the HistoCartography methodology, we introduce a post-hoc graph explainer module to produce compact per-instance explanations emphasizing any diagnostically important entities in the graph. The explainer module can accommodate any graph explainability technique and can perform qualitative and quantitative assessment of the graph explanations generated by the explainability technique. The proposed quantitative assessment results in a set of user-independent quantitative metrics using the statistics of class separability to decipher the diagnostic focus of an AI solution and express the analysis in pathologically understandable terminologies. We test and validate the explainability results using multiple explainer technologies based on inputs from pathologists from three different institutions.

We herein limit our evaluation to cells and cellular interactions via a CG representation in breast cancer subtyping. However, the proposed explainability technique and evaluation framework can be seamlessly adapted to other topological analysis in computational pathology. To this end, we argue that the HistoCartography methodology is a step in the right direction toward improved and interpretable tissue representation and analysis in computational pathology. In the future, we plan to extend our methodology to other biological entities, tissue types, pathological tasks, and imaging modalities. Ultimately, our goal is to understand any information additional to an AI model prediction that one may provide to a clinical user, to build trust and facilitate adoption and deployment of such AI technologies in clinical settings.

Acknowledgments

We are immensely grateful to Maurizio Do Bonito and Gerardo Botti from the National Cancer Institute — IRCCS Fondazione Pascale, Italy, and Giuseppe De Pietro from the Institute for High Performance Computing and Networking — CNR, Italy, for their support and collaborative research with IBM Research Europe.

References

[1] Allemani, C. *et al.* (2015). Global surveillance of cancer survival 1995–2009: Analysis of individual data for 25 676 887 patients from 279 population-based registries in 67 countries (concord-2). *Lancet* 385(9972): 977–1010, doi: 10.1016/S0140-6736(14)62038-9.

[2] Ginsburg, O. *et al.* (2020). Breast cancer early detection: A phased approach to implementation. *Cancer* 126(S10): 2379–2393, doi: 10.1002/cncr.32887.

[3] Pashayan, N. *et al.* (2020). Personalized early detection and prevention of breast cancer: ENVISION consensus statement. *Nat. Rev. Clin. Oncol.* 17: 687–705, doi: 10.1038/s41571-020-0388-9.

[4] Myers, D. J. and Walls, A. L. (2020). Atypical breast hyperplasia. In *StatPearls*, StatPearls Publishing.

[5] Elmore, J. G. *et al.* (2015). Diagnostic concordance among pathologists interpreting breast biopsy specimens. *JAMA* 313(11): 1122–1132, American Medical Association, United States, doi: 10.1001/jama.2015.1405.

[6] Gomes, D. *et al.* (2014). Inter-observer variability between general pathologists and a specialist in breast pathology in the diagnosis of lobular neoplasia, columnar cell lesions, atypical ductal hyperplasia and ductal carcinoma in situ of the breast. *Diagn. Pathol.* 9: 121, doi: 10.1186/1746-1596-9-121.

[7] Allison, K. *et al.* (2014). Understanding diagnostic variability in breast pathology: Lessons learned from an expert consensus review panel. *Histopathology* 65(2): 240–251, doi: 10.1111/his.12387.

[8] Siegel, R. L. *et al.* (2020). Cancer statistics, 2020. *CA Cancer J. Clin.* 70: 7–30, doi: 10.3322/caac.21590.

[9] Mukhopadhyay, S. *et al.* (2017). Whole slide imaging versus microscopy for primary diagnosis in surgical pathology: A multicenter blinded random-ized noninferiority study of 1992 cases (pivotal study). *Am. J. Surg. Pathol.* 42(1): 39–52, doi: 10.1097/PAS.0000000000000948.

[10] Wilbur, D. C. *et al.* (2009). Whole-slide imaging digital pathology as a platform for teleconsultation: A pilot study using paired subspecialist cor-relations. *Arch. Pathol. Lab. Med.* 133(12): 1949–1953, doi: 10.1043/1543-2165-133.12.1949.

[11] Hamilton, P. W. *et al.* (2012). Virtual microscopy and digital pathology in training and education. *APMIS* 120(4): 305–315, doi: 10.1111/j.1600-0463.2011.02869.x.

[12] Sagun, L. and Arias, R. (2018). Digital pathology: An innovative approach to medical education. *Philipp. J. Pathol.* 3(2).

[13] Pantanowitz, L. *et al.* (2018). Twenty years of digital pathology: An overview of the road travelled, what is on the horizon, and the emergence

of vendor-neutral archives. *J. Pathol. Inform.* 9(40), doi: 10.4103/jpi. jpi_69_18.

[14] Nauhria, S. and Hangfu, L. (2019). Virtual microscopy enhances the reliability and validity in histopathology curriculum: Practical guidelines. *MedEdPublish* 8(1): 28, doi: 10.15694/mep.2019.000028.2.

[15] Hanna, M. G. et al. (2019). Whole slide imaging equivalency and efficiency study: Experience at a large academic center. *Mod. Pathol.* 32: 916–928, doi: 10.1038/s41379-019-0205-0.

[16] Litjens, G. *et al.* (2017). A survey on deep learning in medical image analysis. *Med. Image Anal.* 42: 60–88, doi: 10.1016/j.media.2017.07.005.

[17] Liu, Y. *et al.* (2019). How to read articles that use machine learning: Users' guides to the medical literature. *JAMA* 322(18): 1806–1816, doi: 10.1001/jama.2019.16489.

[18] Deng, S. *et al.* (2020). Deep learning in digital pathology image analysis: A survey. *Front Med.* 14(4): 470–487, doi: 10.1007/s11684-020-0782-9.

[19] Sirinukunwattana, K. *et al.* (2016). Locality sensitive deep learning for detection and classification of nuclei in routine colon cancer histology images. *IEEE Trans. Med. Imaging* 35(5): 1196–1206, doi: 10.1109/TMI.2016.2525803.

[20] Höfener, H. *et al.* (2018). Deep learning nuclei detection: A simple approach can deliver state-of-the-art results. *Comput. Med. Imaging Graph.* 70: 43–52, doi: 10.1016/j.compmedimag.2018.08.010.

[21] Kumar, N. *et al.* (2020). A multi-organ nucleus segmentation challenge. *IEEE Trans. Med. Imaging* 39(5): 1380–1391, doi: 10.1109/TMI.2019.2947628.

[22] Graham, S. *et al.* (2019). Hover-Net: Simultaneous segmentation and classification of nuclei in multi-tissue histology images. *Med. Image Anal.* 58: 101563, doi: 10.1016/j.media.2019.101563.

[23] Verma, R. *et al.* (2020). MoNuSAC2020: A Multi-organ nuclei segmentation and classification challenge. *IEEE Trans. Med. Imaging*, doi: 10.1109/TMI.2021.3085712.

[24] Pati, P. *et al.* (2021). Reducing annotation effort in digital pathology: A co-representation learning framework for classification tasks. *Med. Image Anal.* 67: 101859, Elsevier, Netherlands, doi: 10.1016/j. media.2020.101859.

[25] Graham, S. *et al.* (2019). MILD-Net: Minimal information loss dilated network for gland instance segmentation in colon histology images. *Med. Image Anal.* 52: 199–211, doi: 10.1016/j.media.2018.12.001.

[26] Binder, T. *et al.* (2019). Multi-organ gland segmentation using deep learning. *Front. Med.* 6: 173, doi: 10.3389/fmed.2019.00173.

[27] Aresta, G. *et al.* (2019). BACH: Grand challenge on breast cancer histology images. *Med. Image Anal.* 56: 122–139, doi: 10.1016/j.media.2019. 05.010.

[28] Bejnordi, B. E. *et al.* (2017). Diagnostic assessment of deep learning algorithms for detection of lymph node metastases in women with breast cancer. *JAMA* 318(22): 2199–2210, doi: 10.1001/jama.2017.14585.

[29] Pati, P. *et al.* (2018). Deep positive-unlabeled learning for region of interest localization in breast tissue images. In *SPIE Medical Imaging 2018: Digital Pathology*, Vol. 10581, pp. 1058107, doi: 10.1117/12.2293721.

[30] Veta, M. *et al.* (2019). Predicting breast tumor proliferation from wholeslide images: The TUPAC16 challenge. *Med. Image Anal.* 54: 111–121, doi: 10.1016/j.media.2019.02.012. Elsevier, Netherlands.

[31] Noorbakhsh, J. *et al.* (2020). Deep learning-based cross-classifications reveal conserved spatial behaviors within tumor histological images. *Nat. Commun.* 11(6367), doi: doi.org/10.1038/s41467-020-20030-5.

[32] Campanella, G. *et al.* (2019). Clinical-grade computational pathology using weakly supervised deep learning on whole slide images. *Nat. Med.* 25(8): 1301–1309, doi: 10.1038/s41591-019-0508-1.

[33] Bulten, W. *et al.* (2020). Automated deep-learning system for Gleason grading of prostate cancer using biopsies: A diagnostic study. *Lancet Oncol.* 21(2): 233–241, doi: doi.org/10.1016/S1470-2045(19)30739-9.

[34] Mercan, C. *et al.* (2019). From patch-level to ROI-level deep feature representations for breast histopathology classification. In *SPIE Medical Imaging 2019: Digital Pathology*, Vol. 10956, pp. 109560H, doi: 10.1117/12.2510665.

[35] Qaiser, T. *et al.* (2017). Her2 challenge contest: A detailed assessment of automated Her2 scoring algorithms in whole slide images of breast cancer tissues. *Histopathology* 72(2): 227–238, Wiley, Hoboken, New Jersey, United States, doi: 10.1111/his.13333.

[36] Kashyap, A. *et al.* (2019). Quantitative microimmunohistochemistry (qμIC): A method to grade immunostains in tumor tissues using saturation kinetics. *Nat. Biomed. Eng.* 3: 478–490, doi: 10.1038/s41551-019-0386-3.

[37] Kaigala, G. *et al.* Biomarker quantification in a tissue sample. U.S. Patent 2019/0286790 A1, issued September 19, 2019.

[38] Arar, N. M. *et al.* (2019). High-quality immunohistochemical stains through computational assay parameter optimization. *IEEE Trans. Biomed. Eng.* 66(10): 2952–2963, doi: 10.1109/TBME.2019.2899156.

[39] Arar, N.M. *et al.* Tissue staining quality determination. U.S. Patent 10,706,535 B2, issued March 14, 2019.

[40] Bejnordi, B. E. *et al.* (2016). Stain specific standardization of whole-slide histopathological images. *IEEE Trans. Med. Imaging* 35(2): 404–415, doi: 10.1109/TMI.2015.2476509.

[41] Vahadane, A. *et al.* (2016). Structure-preserving color normalization and sparse stain separation for histological images. *IEEE Trans. Med. Imaging* 35(8): 1962–1971, doi: 10.1109/TMI.2016.2529665.

[42] Stanisavljevic, M. *et al.* (2019). A fast and scalable pipeline for stain normalization of whole-slide images in histopathology. In *European Conference on Computer Vision Workshops*, Vol. 11134, doi: 10.1007/978-3-030-11024-6_32.

[43] Mobadersany, P. *et al.* (2018). Predicting cancer outcomes from histology and genomics using convolutional networks. *PNAS*, 115(13), United States National Academy of Sciences, Washington, D.C., United States, doi: 10.1073/pnas.1717139115.

[44] Wulczyn, E. *et al.* (2020). Deep learning-based survival prediction for multiple cancer types using histopathology images. *PLOS ONE*, doi: 10.1371/journal.pone.0233678.

[45] Bhargava, R. and Madabhushi, A. (2016). Emerging themes in image informatics and molecular analysis for digital pathology. *Annu. Rev. Biomed. Eng.* 18: 387–412, doi: 10.1146/annurev-bioeng-112415-114722.

[46] Wang, S. *et al.* (2019). Artificial intelligence in lung cancer pathology image analysis. *Cancers* 11(11): 1673, doi: 10.3390/cancers11111673.

[47] Ibrahim, A. *et al.* (2020). Artificial intelligence in digital breast pathology: Techniques and applications. *Breast* 49: 267–273, doi: 10.1016/j.breast.2019.12.007.

[48] Pantanowitz, L. *et al.* (2020). An artificial intelligence algorithm for prostate cancer diagnosis in whole slide images of core needle biopsies: A blinded clinical validation and deployment study. *Lancet Digital Health* 2(8): E407–E416, doi: 10.1016/S2589-7500(20)30159-X.

[49] Rathore, S. *et al.* (2019). Segmentation and grade prediction of colon cancer digital pathology images across multiple institutions. *Cancers* 11(11): 1700, doi: 10.3390/cancers11111700.

[50] Song, Z. *et al.* (2020). Clinically applicable histopathological diagnosis system for gastric cancer detection using deep learning. *Nat. Commun.* 11(4294), doi: 10.1038/s41467-020-18147-8.

[51] Kiani, A. *et al.* (2020). Impact of a deep learning assistant on the histopathologic classification of liver cancer. *NPJ Digit. Med.* 3(23), doi: 10.1038/s41746-020-0232-8.

[52] Bejnordi, B. E. *et al.* (2017). Context-aware stacked convolutional neural networks for classification of breast carcinomas in whole-slide histopathology images. *J. Med. Imaging* 4(4): 044504, doi: 10.1117/1.JMI.4.4.044504.

[53] Sirinukunwattana, K. *et al.* (2018). Improving whole slide segmentation through visual context — A systematic study. *Med. Image Comput. Comput. Assist. Interv.* 11071, doi: 10.1007/978-3-030-00934-2_22.

[54] Tellez, D. *et al.* (2019). Neural image compression for gigapixel histopathology image analysis. *IEEE Trans. Pattern Anal. Mach. Intell.* 43(2): 567–578, doi: 10.1109/TPAMI.2019.2936841.

[55] Gartner, L. (2020). Textbook of histology (5th ed.). Elsevier, Netherlands.

[56] Bruno, K. *et al.* (2017). Looking under the hood deep neural network visualization to interpret whole slide image analysis outcomes for colorectal polyps. In *IEEE Conference on Computer Vision and Pattern Recognition Workshops*, pp. 821–827, doi: 10.1109/CVPRW.2017.114.

[57] Binder, A. *et al.* (2018). Towards computational fluorescence microscopy: Machine learning-based integrated prediction of morphological and molecular tumor profiles. in arXiv:1805.11178.

[58] Graziani, M., Andrearczyk, V., Marhcand-Maillet, S. and Muller, H. (2020). Concept attribution: Explaining CNN decisions to physicians. *Comput. Biol. Med.* 123: 103865, Elsevier, Netherlands, doi: 10.1016/j.compbiomed.2020.103865. Elsevier, Netherlands.

[59] Lu, M. Y. *et al.* (2021). Data efficient and weakly supervised computational pathology on whole slide images. *Nat. Biomed. Eng.*, doi: 10.1038/s41551-020-00682-w.

[60] Zhang, Z. *et al.* (2019). Pathologist-level interpretable whole-slide cancer diagnosis with deep learning. *Nat. Mach. Intell.* 1: 236–245, doi: 10.1038/s42256-019-0052-1.

[61] Demir, C. G. *et al.* (2004). The cell graphs of cancer. *Bioinformatics* 4(20): 145–151, doi: 10.1093/bioinformatics/bth933.

[62] Sharma, H. *et al.* (2017). A comparative study of cell nuclei attributed relational graphs for knowledge description and categorization in histopathological gastric cancer whole slide images. In *IEEE Symposium on Computer-Based Medical Systems*, pp. 61–66, doi: 10.1109/CBMS.2017.25.

[63] Zhou, Y. et al. (2019). CGC-Net: Cell graph convolutional network for grading of colorectal cancer histology images. *IEEE/CVF International Conference on Computer Vision Workshops*, IEEE, Long Beach, CA, United States, doi: 10.1109/ICCVW.2019.00050.

[64] Pati, P. et al. (2020). HACT-Net: A hierarchical cell-to-tissue graph neural network for histopathological image classification. *Uncertainty for Safe Utilization of Machine Learning in Medical Imaging, and Graphs in Biomedical Image Analysis, UNSURE 2020, GRAIL 2020*, Vol. 12443, pp. 208–219, Springer, Lima, Peru, doi: 10.1007/978-3-030-60365-6_20.

[65] Chen, R. J. *et al.* (2020). Pathomic fusion: An integrated framework for fusing histopathology and genomic features for cancer diagnosis and prognosis. *IEEE Trans. Med. Imaging.*, doi: 10.1109/TMI.2020.3021387.

[66] Anand, D. *et al.* (2020). Histographs: Graphs in histopathology. In *SPIE Medical Imaging 2020: Digital Pathology*, Vol. 11320, pp. 113200O, doi: 10.1117/12.2550114.

[67] Sharma, H. et al. (2015). A review of graph-based methods for image analysis in digital histopathology. *Diagn. Pathol.*, doi: 10.17629/www.diagnosticpathology.eu-2015-1:61.

[68] Aygunes, B. *et al*. (2020). Graph convolutional networks for region of interest classification in breast histopathology. *SPIE Medical Imaging 2020: Digital Pathology*, Vol. 11320, pp. 113200K, doi: 10.1117/12.2550636. SPIE, San Diego, CA, United States.

[69] Javed, S. *et al*. (2020). Cellular community detection for tissue phenotyping in colorectal cancer histology images. *Med. Image Anal.* 63: 101696, Elsevier, Netherlands, doi: 10.1016/j.media.2020.101696.

[70] Zhao, Y. *et al*. (2020). Predicting lymph node metastasis using histopathological images based on multiple instance learning with deep graph convolution. In *IEEE/CVF Conference on Computer Vision and Pattern Recognition*, pp. 4837–4846, doi: 10.1109/CVPR42600.2020.00489.

[71] Adnan, M. *et al*. (2020). Representation learning of histopathology images. using graph neural networks. *IEEE/CVF Conference on Computer Vision and Pattern Recognition Workshops*, pp. 4254–4261, doi: 10.1109/CVPRW50498.2020.00502. IEEE.

[72] Jaume, G. *et al*. (2020). Towards explainable graph representations in digital pathology. In *International Conference on Machine Learning Workshops*.

[73] Sureka, M. *et al*. (2020). Visualization for histopathology images using graph convolutional neural networks. *IEEE International Conference on Bioinformatics and Bioengineering*, pp. 331–335, doi: 10.1109/BIBE50027.2020.00060. IEEE, Cincinnati, OH, United States.

[74] Holzinger, B. *et al*. (2017). Towards the augmented pathologist: Challenges of explainable-ai in digital pathology. in: arXiv:1712.06657.

[75] Kumar, N. *et al*. (2017). A dataset and a technique for generalized nuclear segmentation for computational pathology. *IEEE Trans. Med. Imaging* 36(7): 1550–1560, doi: 10.1109/TMI.2017.2677499.

[76] Graziani, M. *et al*. (2018). Regression concept vectors for bidirectional explanations in histopathology. In *Understanding and Interpreting Machine Learning in Medical Image Computing Applications*, Vol. 11038, doi: 10.1007/978-3-030-02628-8_14.

[77] Hägele, M. *et al*. (2020). Resolving challenges in deep learning-based analyses of histopathological images using explanation methods. *Sci. Rep.* 10: 6423, Nature, Berlin, Germany, doi: 10.1038/s41598-020-62724-2.

[78] Bach, A. *et al*. (2015). On pixel-wise explanations for non-linear classifier decisions by layer-wise relevance propagation. *PLoS ONE*, 10(7), doi: 10.1371/journal.pone.0130140.

[79] Jaume, G. *et al*. (2021). Quantifying explainers of graph neural networks in computational pathology. *IEEE/CVF Conference on Computer Vision and Pattern Recognition*, pp. 8106-8116.

[80] Ying, R. *et al*. (2019). GNNExplainer: Generating explanations for graph neural networks. *Proceedings of the 33rd International Conference on*

Neural Information Processing Systems, 9240–9251. Curran Associates Inc., Vancouver, Canada.

[81] Pope, P. E. *et al.* (2019). Explainability methods for graph convolutional neural networks. *IEEE/CVF Conference on Computer Vision and Pattern Recognition*, pp. 10764–10773, IEEE, Long Beach, CA, United States, doi: 10.1109/CVPR.2019.01103.

[82] Schwarzenberg, R. *et al.* (2019). Layerwise relevance visualization in convolutional text graph classifiers. *Proceedings of the Thirteenth Workshop on Graph-Based Methods for Natural Language Processing*, pp. 58–62, Association for Computational Linguistics, Hong Kong, doi: 10.18653/v1/D19-5308.

[83] Macenko, M. *et al.* (2009). A method for normalizing histology slides for quantitative analysis. *IEEE International Symposium on Biomedical Imaging: From Nano to Macro*, pp. 1107–1110, IEEE, Boston, Massachusetts, United States, doi: 10.1109/ISBI.2009.5193250.

[84] Achanta, R. *et al.* (2012). SLIC superpixels compared to state-of-the-art superpixel methods. *IEEE Trans. Pattern Anal. Mach. Intell.*, 34(11): 2274–2282, doi: 10.1109/TPAMI.2012.120. IEEE.

[85] Potjer, F. (1996). Region adjacency graphs and connected morphological operators. *Mathematical Morphology and its Applications to Image and Signal Processing. Computational Imaging and Vision*, 5: 111–118, Springer, doi: 10.1007/978-1-4613-0469-2_13.

[86] Gilmer, J. *et al.* (2017). Neural message passing for quantum chemistry. *Proceedings of the 34th International Conference on Machine Learning*, 70: 1263–1272. Journal of Machine Learning Research, Sydney, Australia.

[87] Defferrard, M. *et al.* (2016). Convolutional neural networks on graphs with fast localized spectral filtering. Proceedings of the 30th International *Conference on Neural Information Processing Systems*, 3844–3852. Curran Associates Inc., Barcelona, Spain.

[88] Hamilton, W. *et al.* (2017). Inductive representation learning on large graphs. *Proceedings of the 31st International Conference on Neural Information Processing Systems*, 1025–1035. Curran Associates Inc., Long Beach, CA, United States.

[89] Kipf, T. N. and Welling, M. (2017). Semi-supervised classification with graph convolutional networks. *International Conference on Learning Representations*, Toulon, France.

[90] Velickovic, P. *et al.* (2018). Graph attention networks. *International Conference on Learning Representations*, Vancouver, Canada.

[91] Xu, K. *et al.* (2019). How powerful are graph neural networks? *International Conference on Learning Representations*, New Orleans, Louisiana, United States.

[92] Corso, G. *et al.* (2020). Principal neighborhood aggregation for graph nets. *Proceedings of the 34th International Conference on Neural Information Processing Systems.*

[93] Dwivedi, V.P. *et al.* (2020). Benchmarking graph neural networks. arXiv:2003.00982.

[94] Xu, K. *et al.* (2018). Representation learning on graphs with jumping knowledge networks. *Proceedings of the 35th International Conference on Machine Learning* 80: 5453–5462. Proceedings of Machine Learning Research, Stockholm, Sweden.

[95] Kanan, C. et al. (2020). Independent validation of paige prostate: Assessing clinical benefit of an artificial intelligence tool within a digital diagnostic pathology laboratory workflow. *J. Clin. Oncol.* 38(15), doi: 10.1200/ JCO.2020.38.15_suppl.e14076. American Society of Clinical Oncology, Alexandria, Virginia, United States.

[96] Ström, P. et al. (2020). Artificial intelligence for diagnosing and grading of prostate cancer in biopsies: a population-based, diagnostic study. *Lancet Oncol.* 21(2): 222–232, doi: 10.1016/S1470-2045(19)30738-7.

[97] Esteva, A. *et al.* (2017). Dermatologist-level classification of skin cancer with deep neural networks. *Nature* 542: 115–118, doi: 10.1038/nature21056. Nature, Berlin, Germany.

[98] Coudray, N. *et al.* (2018). Classification and mutation prediction from non–small cell lung cancer histopathology images using deep learning. *Nat. Med.* 24: 1559–1567, doi: 10.1038/s41591-018-0177-5.

Chapter 11

AI-Driven Design of Disease Sensors: Theoretical Foundations

Simone Bianco[*,†], **Sara Capponi**[*,‡], **and Shangying Wang**[*,§]

IBM Almaden Research Center, San Jose, California, USA
NSF Center for Cellular Construction, San Francisco, California, USA

†*sbianco@us.ibm.com*
‡*sara.capponi@ibm.com*
§*swang@ibm.com*

1. Introduction

Over the last decade, artificial intelligence (AI) and machine learning (ML) have become integral to biological analytics, with applications ranging from image analysis [1–3], genome annotation [4, 5], to predictions of protein structure and binding [6, 7], cancer transcriptomics [8, 9], predictions of metabolic functions in complex microbial communities [10], and the characterization of transcriptional regulatory networks [11, 12]. To date, however, the intersection of AI and synthetic biology has been limited, with few notable exceptions like the field of RNA [10, 13] or protein design [14], a small number of custom-made solutions in metabolic

---

*All authors contributed equally to the work.

engineering [15, 16] and in T cell engineering [17, 18]. Therefore, the enormous potential of AI to design novel biological functions is yet to be fulfilled.

Synthetic biology seeks to use biology as an engineering substrate to program novel user-defined functions with molecules, cells, and biomaterials. Advancements in this field have resulted in the development of living cells as production factories for high-value small molecules and proteins [19], pharmaceuticals [20–22], food additives [23], feedstock and raw materials [24], or biopolymers [25]. Despite many successes, and the immense promise of biology as a source of technological innovation, progress has been slow. It is still an unfortunate reality that circuits built with biological molecules rarely function as intended and that their design and refinement typically requires many months and even years of trial and error. To put it in perspective, it took 150 person-years of effort for heterologous expression of the 16-enzyme artemisinin pathway [22], and 575 person-years of effort for DuPont's 1,3-propanediol [26]. Many factors hinder progress. First, our understanding of core design principles for biological circuits remains very limited. Second, our toolbox of diverse, well-characterized components for circuit construction is limited. Third, we still have little understanding of how biological regulatory processes connect across scales, from sequences to cells and cell collectives. Finally, we do not know how to leverage prior biological knowledge (genetic, structural, and biophysical studies that span decades) to power empirical approaches (e.g., using AI/ML) for biological design. As synthetic biology extends its reach into broad application areas (e.g., health, agriculture, energy, environment), there is a growing need to take on these challenges so as to make biological design more predictable and time-efficient. AI approaches are particularly well suited for this task, mostly because of their ability to learn complex patterns from data and to be predictive without the need for detailed mechanistic understanding. The use of AI has therefore tremendous potential for accelerating the engineering of biology.

The field of artificial cellular sensing represents a recent and exciting avenue of research in the broad field of synthetic biology [27]. Cellular sensors, i.e., artificially modified cells that act as reporters of specific chemical, biological, and physical perturbations, are desirable for a variety of applications where minimal disruption of the biological environment is required. Yet, as mentioned above, their design is left to expensive trial and error or serendipitous discoveries. In this chapter, we lay the theoretical foundations for a general AI-driven platform for the design of cellular sensors with specific functionalities. We are particularly interested in disease biosensors, that is biological machines capable of reporting on the presence, onset, or progression of a given disease [28–30]. Our aim is not to practically demonstrate the design of a complex sensor. While we believe that task to be within reach in this generation, its practical realization is beyond our technical knowledge. Rather, we want to describe the technical challenges and provide ideas to advance the field.

2. Constraints of Computational Cellular Engineering

Designing synthetic cells for specific purposes is experimentally within reach. Advances in gene editing like CRISPR-CAS9, as well as the development of artificial nano-containers and transport platforms has accelerated the realization of artificial biological structures and even whole cells [31]. Two main experimental strategies exist: A bottom-up approach, which uses forward DNA engineering to realize minimal genomes capable of encoding for interesting products [31], and a top-down approach, where the synthetic product is the result of removing superfluous genes from an existing genome, to finally realize the desired product [32]. It is evident that these two approaches are still in the realm of trial and error. Nevertheless, synthetic cells are routinely realized.

From a computational perspective, there are very few rigorous design platforms for synthetic biology. Most computational systems are limited to the design of gene regulatory networks [33, 34]. Some progress has been made in the design of complex microbial colonies, as in the case of the Automated Recommendation Tool for the computational design of microbiomes with assigned collective functions [15]. The lack of appropriate design tools can be ascribed to a number of essential knowledge gaps in the understanding of the process of biological self-assembly. First of all, cellular assembly is a *multi-scale process*. From the genome to the cell and cell collectives, biological assembly is a process spanning several spatial and time scales. Moreover, biological entities at all scales (from pathways to whole cells) have been tuned by billions of years of evolution to perform functions necessary for the cell to maintain homeostasis. Thus, the modifications of essential and nonessential biological processes, whether they be at genetic, molecular, cellular, or tissues/

organism level, may have unexpected consequences and result in the production of unanticipated products or the shutdown of other essential avenues. Thirdly, the mechanisms regulating the assembly of biological entities are not completely known. The relationship between genome sequences, biochemical structures, and cellular functions is not considered entirely predictable, thus adding a randomness which is not conducive to the development of a design platform. Strictly connected with the previous point is the unique role of noise in biological systems. To this extent, an interesting analogy can be made for the fields of physics and engineering. Both disciplines have arguably dominated the science of the 20th century. It has been said many times that the 21st century will be the century of biology. Yet, when thinking about the role of noise in biological systems, a lot is still unclear. Physics and engineering have devoted entire disciplines to the study of noise and its properties, namely statistical physics and control engineering, respectively. More specifically, control engineering has developed very effective theoretical and practical methods to reduce the impact of noise on physical systems. This is a feature whose importance would be impossible to overstate in fields like aerospace engineering and civil engineering. In those contexts, noise is a variable that, if not accounted for, may produce unexpected and catastrophic effects. Though noise is generally treated as the opposite of information, it is also understood as a source of novelty and variation in biology [35–38]. Noise can propagate through a genetic cascade and play important functional roles [39–42]. In fact, random variations in the numbers of specific chemical products may result in the activation of a particular pathway instead of another. In a sense, we could say that noise is not a "bug", but it is a "feature" of biological systems. As such, it needs to be understood in the context of a living system with a "purpose", and appropriately considered when design principles are encoded and utilized. Which brings us to the last point: Evolution. The famous sentence that "Nothing in biology makes sense except in the light of evolution" [43] is even more true in the case of artificial systems. In fact, a cell or any other biological system has evolved to survive and reproduce in a given environment (or ensemble of environments). Yet, synthetic biology aims at subverting the innate function of a biological system, forcing it to produce products that are important or interesting for the investigator. Therefore, a certain degree of uncertainty about cellular fate, resilience to perturbation, and robustness is to be expected when a biological system is artificially altered.

3. A Multi-Scale Problem

Cellular self-assembly is a multi-scale problem, bridging the intracellular molecular dynamic, network biology, and gene regulation, with the whole-cell organization and the cell to cell interactions. In order for an AI to generate design principles of cellular sensors, a precise understanding of cellular sensing machinery needs to be obtained at all levels. As Endy theorizes in a landmark review paper published in Nature [44], to design a biological system requires a certain abstraction of the biological process, which suits the particular investigator or engineer. This view of synthetic biology is important, since it means that an investigator interested in a physiological output need not know the inner workings of the biological system to be able to engineer it. This tendency to simplify the bioengineering process stems directly from a standardization of the field, whereas parts, components, rules of assembly are well codified and do not require to be invented every time a new product is needed. One can imagine that a bioengineer of the future will not need to know intimately the rules of DNA binding or assembly in order to generate a given product, but only to know which "parts" to order so that the final product can be built following well-established rules and protocols. In some sense, the field is striving to become closer to engineering in nature, with a higher control of the end product and a mechanistic predictability of the outcome. To achieve this ambitious goal, a computational framework, akin to a CAD software, is necessary. Computation in biology plays an important role in generalizing and elucidating the complications of cellular processes by abstracting them, when possible, into sets of equations and algorithms. The explosion of systems biology, often associated with synthetic biology, has provided a strong quantitative foundation for the mechanistic design of biological structures, from biochemical networks to cells and tissues. Indeed, it has become common to computationally explore a large parameter space, generating in silico thousands of alternative scenarios, which would take often years and millions of dollars to be tested experimentally. Today, the rational design of cells and cellular structures is a solvable interdisciplinary multi-scale problem, and part of its solutions routinely involves modeling mutated pathways to infer potential design principles of cells. Yet, the rational design of cells is still elusive, and the way they are engineered is still *ad hoc* and relatively limited in scope. Rather, the myriad of studies on the subject published in the past few decades paint a complex and interconnected picture of the multi-scale dynamic of a cell.

Although computational biology is fundamental to design principles and making predictions, an approach based solely on mechanistic models is bound to suffer from many drawbacks. Abstracting a biological process provides many times a limited relation to the experimental world. Model parameters are not always related to specific biological quantities, and therefore may be difficult to alter or control experimentally. The overall network biology may not be known and the causal inference between links may be insufficient to paint a complete picture. Moreover, while the relevant biochemistry may be modeled as a local process, there are overall changes that happen post-modification in the cell because of the complexities involved in the regulation of cell fate and homeostasis. Additionally, bridging scales requires non-trivial computational resources [45]. Finally, while whole-cell models are possible [46], their complications may make them extremely difficult to generalize. Synthetic cellular design requires a paradigm shift, moving away from modeling only the biochemistry of the cell, to looking at principles of whole-cell organization instead [47]. This would be akin to a chemical engineering factory being built not only by considering the chemistry necessary for the product, but also and importantly by tuning type, size, and organization of the reaction vessels. Innovation in this field needs to come from the realization that the process of self-assembly is exquisitely multi-scale. AI and ML can help us address some of these shortcomings. ML can help uncover systems' properties by inferring relationships directly from data. Thus, it should not be surprising to consider the first approach to the biological design to be one bridging direct mathematical modeling of cells and cellular pathways with AI-infused data-driven models.

4. Design Principle for Engineering Biosensors

The field of synthetic sensor engineering has produced a number of notable advances [48–52]. In this section, we will introduce a prototypical example of an engineered biosensor. A very promising recent platform for the realization of multi-cellular structures with desired properties is the synNotch (from synthetic Notch) signaling platform, invented by the Lim lab at UC San Francisco [53]. These are minimal engineered intracellular platforms linked to a chimeric extracellular recognition domain and a chimeric intracellular transcriptional domain. When the external domain is recognized by an external cell, the synNotch activates the internal pathway, driving targeted gene expression. The power of synNotch and other similar artificial programs is that they are capable of selectively activating expression upon specific external perturbations, which makes them amenable to being used as cellular sensors. For example, Toda and coworkers have been able to engineer synNotch equipped cells with fluorescence reporters which activate when some physical conditions are met [53].

Precise cell–cell communication is important in multi-cellular organisms, such as in the brain, the developing embryo, and the immune system. Cells can precisely sense the environment, recognize one another, and in general achieve complex behavior to maintain a functional organism. Disease tend to disrupt this delicate order. In cancer, the patient's own immune systems, T cells specifically, are unable to recognize cancer cells, thus failing to trigger an effective immune response against the disease. One of the most exciting developments of the last decade has been to use chimeric antigen receptors (CAR), which effectively are synthetic sensors, which are capable of reprogramming T cells to attack cancer cells [54, 55]. The broad applicability of this kind of therapy is limited by the high specificity needed to target tumor-only antigens [56, 57].

The rational design of sensors and their related pathways to autonomously sense user-specified disease or injury signals, and precisely deploying therapeutic or repair functions, is critical to the success of CAR-T therapy [58]. Tasks that once seemed infeasible such as designing and implementing intricate synthetic gene circuits that perform complex sensing and actuation functions are now being realized. These achievements have been made possible by the integration of diverse expertise across biology, physics, and engineering, resulting in an emerging, quantitative understanding of biological design.

The notch receptor, a classic receptor in developmental biology, can mediate direct cell–cell communication [59]. It is composed of three parts: an extracellular ligand-binding domain (sensing), a regulatory domain (signal transduction), and the intracellular effector domain (response). The synNotch receptor was created by removing the sensing and response domains of the native Notch receptor, and replacing them with different recognition and response domains to be able to change what extracellular contact ligands the cell detects, what intracellular transcriptional regulator is used, and what intracellular effector genes this regulator drives upon engagement. The sensing and response domains of the synNotch receptor can be flexibly altered in a modular fashion leading to highly customizable sensing/response pathways. The wide variety of possible synNotch receptor architectures and domain combinations yield powerful and flexible tools for engineering cell–cell communication [60]. This modular process is amenable to ML and translatable to machine intelligence.

When designing a cellular receptor, it is also important to make sure it can function orthogonal to other receptors. Orthogonal function is the key for a cell to elaborate different outputs according to the presence or absence of multiple inputs, allowing for combinatorial input integration from multiple receptors with little crosstalk. This ensures that synthetic receptors can work independently and compatibly.

By incorporating synNotch receptors into CAR-T cells, sensing and killing cancer cells can be done in a more precise and reliable way. Activation of a synNotch receptor for one antigen drives the inducible expression of a CAR for a second antigen. These dual-receptor T cells are only armed and activated in the presence of dual-antigen tumor cells (AND-gate). Only after detecting the first antigen on cancer cells will the cells become armed for killing, making it neutral to bystander cells, thus preventing lethal side effects, leading to precise immune recognition of a much larger set of tumors [61].

SynNotch provides an extraordinarily flexibile platform for engineering mammalian cells with customized sensing/response behaviors to achieve combinatorial integration of user-specified environmental cues. Thus, one can build a receptor targeted to a cell surface ligand of interest, such as a disease-related or tissue-related antigen, and this environmental sensing event leads to the release of the transcriptional regulator and the initiation of a custom cellular response. SynNotch receptors can be used in a wide variety of cell types to help them sense their environment and locally modulate their own behavior or the surrounding microenvironment [60–63]. SynNotch receptor is also a good demonstration on how a successful biosensor can be engineered: with modularity, compatibility, predictability, and reliability.

5. Integrating AI with Mathematical Modeling

It is often challenging to convert design concepts to predicted results in synthetic biology. Not only because of the nonlinear and stochastic character of biological systems, but also because of many tunable parameters (such as strength of interactions, transcription rate, etc.), which can lead to a variety of distinct outputs even for exactly the same design. This stumbling block can be alleviated by the use of computer-aided mathematical modeling. Mathematical models have been a powerful and often indispensable tool for guiding the design and implementation of synthetic patterns and for experimental data interpretation and validation [64–69]. They can help with the identification of the regions of parameter space that produce the desired behavior or the most effective design by searching large parameter spaces *in silico*. They also provide the capability of using knowledge about the constituent parts of a system to predict the behavior of a system as a whole. Therefore, mathematical modeling serves as a bridge connecting a conceptual design idea to its biological realization.

Quantitative understandings of a few relatively simple gene regulatory networks, such as positive-feedback networks, negative-feedback networks, and oscillators, not only provide insights into mechanisms underlying behaviors of more complex gene networks, but also lay the foundation to the early success of synthetic biology [70, 71].

Positive-feedback loops or double-negative-feedback loops can, in principle, convert graded inputs into switch-like, irreversible responses. They enable signal amplification, hysteresis, and bistability [72]. Gardner *et al.* engineered the first double-negative-feedback systems, named toggle switch, into E. coli. The system could be made to toggle between the on and off states by the addition of external trigger stimuli [70]. Kramer and Fussenegger designed and characterized a synthetic mammalian gene circuit, consisting of a tetracycline-responsive positive-feedback loop, showing hysteretic signal integration. This engineered system consisted of well-studied components enabling precise mathematical modeling and testing of mechanistic sufficiency. Guided by predictions in silico, hysteresis was observed in experiments with experimental parameters comparable to modeling parameters [73].

The negative-feedback loop has been shown to accelerate transcriptional response time [74] and reduce gene expression noise in bacteria and yeast [75, 76]. However, counter-intuitive theoretical and experimental results show that negative-feedback might not reduce noise, or amplify noise in gene expression in certain scenarios [41, 77, 78]. For negative augoregulation, researchers synthetically constructed the transcriptional negative autoregulatory motif and integrated the circuits stably in human kidney cells. By using this synthetic system, they found that negative-feedback reduces extrinsic noise while it marginally increases intrinsic noise [79].

Oscillatory gene networks combine positive-feedback and negative-feedback modules with a time delay to achieve periodically changing gene expression and/or cellular phenotype [80]. The first synthetically engineered oscillator is the repressilator in *Escherichia coli*. Fung *et al.* reported the mathematical model-aided design of a gene-metabolic oscillatory circuit called the metabolator [81]. Early efforts reconstructed an oscillatory gene network in mammalian cells to display the circadian rhythm [82].

There are many more successful examples of synthetic systems consisting of these simple regulatory circuits. These simple systems are the building blocks and recurrently used in synthetically engineered biological systems, including synthetic cellular sensors. In developing these synthetic systems, mathematical modeling plays an important role in understanding the mechanisms and optimizing parameters for functions. However, for more complex systems, especially for systems that deal with spatial or stochastic dynamics, or for multi-scale systems, much larger computational power is needed, as well as a thorough investigation and validation of the mathematical models. Both the size of the parametric space and the time required to do each simulation would increase combinatorially with the system complexity. Thus, standard numerical simulations using mechanism-based models can face a prohibitive barrier for large-scale exploration of system behaviors.

AI, especially deep learning, can potentially overcome the computational bottleneck faced by many mathematical models. When numerically solving a mechanism-based dynamic model consisting of differential equations, the majority of the time is spent in the generation of time courses. However, for most biological questions, the main objective is to map the input parameters to specific outcomes. For such applications, the time-consuming generation of time courses is a necessary evil. The key to the massive acceleration in predictions is to use the deep learning to bypass the generation of fine details of system dynamics but instead focus on an empirical mapping between input parameters to system outputs of interest.

The growing prevalence of ML approaches offer a newer-age alternative to the classic systems approach. Integrating these advances to establish a next-generation paradigm of synthetic biology will undoubtedly yield meaningful results [69, 83]. The general principle is to use a small proportion of data generated by the mechanism-based model to train a neural network. The data generated by the mechanistic model need to be sufficiently large to ensure reliable training but small enough such that the data generation is computationally feasible [83]. Integrating AI with mathematical modeling has many benefits, including fast exploration of parameter space of known model structure [83, 84], model structure inference [85], and generation of novel biological insights [86].

To date, however, the predicted outputs are restricted in categorical labels or a set of discrete values. By contrast, deep learning has not been used to predict outputs consisting of continuous sequences of data (e.g., time series, spatial distributions, and probability density functions). A pioneer work done by Wang *et al.* has demonstrated that by adopting a special type of deep learning network, the Long-Short-Term Memory (LSTM) network, this limitation can be overcome [83]. In their research, a synthetic gene circuit, accounting for cell growth and movement, intercellular signaling, and circuit dynamics as well as transportation was engineered in *Escherichia coli* [83, 87]. This circuit can generate spatial patterns, i.e., 1-, 2-. 3-ring patterns in bacterial colony growth. However, the data generated in the experiments, as well as in numerical simulations from a mathematical model are sparse in the parameter space in finding the parametrization to generate complex (3-ring) patterns. The input of the neural network can be a variety of variables that can interfere with the system output and the output of the neural network is a 1D spatial distribution due to circular symmetry. By employing the LSTM neural networks, together with ensemble voting strategy, they are able to reveal the general criterion for making complex patterns [83].

Although much of the effort in engineering cellular sensors has been devoted to metabolic and signaling pathways, another aspect of cells that has received far less attention is their physical structure or geometry [47]. Indeed, the relationship between chemical reactions and the morphological features of the reaction vessels containing them is still not completely understood. This is particularly true for bioengineered sensors, where the geometry of organelles and even of the whole cell can be an easy-to-measure readout of the sensing process, which is itself amenable to the adoption of AI for image processing and classification [88–92].

Currently, there are no known methods for the rational engineering of cells with a predictable cellular architecture. However, a lot of attention has been devoted to the cellular surface and how it is affected by the bioengineering process. In particular, advances in the fields of molecular dynamics have generated enormous excitement in synthetic biology. The next section will highlight some recent results.

6. Recent Advances on Use of AI in Molecular Dynamics and Simulations

The first atomistic molecular dynamics simulation of a biological macro-molecule published in 1977 [93] was a scientific breakthrough not only with respect to the description of the dynamics of a protein but also because it demonstrated that a computational approach such as molecular dynamics already used in describing the features of fluids [94] could be employed successfully to examine protein motions. In this seminal paper, the authors simulated in vacuum the bovine pancreatic trypsin inhibitor (BPTI) formed by 58 amino acids for 9.2 ps using a rough molecular mechanistic potential. Since then, the fields of molecular modeling and simulation have expanded thanks to the development of both hardware and software so that calculations can handle a very high number of atoms and can be performed in parallel across multiple processors. Also, the accuracy of force fields, i.e., the computational methods used to estimate and model the interaction potential between atoms and molecules, used during the simulations improved notably allowing more extensive valida-tion of these force fields against experimental data, which led to an expan-sion of the fields of application.

The development of this field was so important that in 2013 the Nobel Prize in Chemistry was assigned among others to Martin Karplus, one of the authors of that seminal paper published in 1977 for "the development of multiscale model for complex chemical systems". After more than 30 years, this Nobel Prize recognizes the importance of the computational approaches to examine nature and natural phenomena revealing the bridges that connect physics, chemistry, mathematics, and computer sci-ence to biology.

Molecular dynamics is now a well-established approach to examine the structure and dynamics of biological systems [95–98]. Simulations provide details at atomistic resolution on single motion as a function of time. In addition to study or validate experimental data, simulations can explore uncharted regions of the biological energy landscape not acces-sible experimentally because the energy potential is defined by the user and so prone to modification. M. Karplus and J. A. McCammon identify three main fields in which molecular dynamics simulations play a major role [98], the first being sampling the configuration space, the second, examining the dynamics and motions of a system at equilibrium, and the third being the study of the dynamics and conformational changes of a macromolecule. At this point, several factors need to be taken into account with respect to the early days in which simulations of several picosecond were considered extraordinary effort. Although atomistic, molecular dynamics simulations can be used to describe motions of systems of 10^5–10^6 atoms spanning from femto second to the order of millisecond and from 10^{-1} to 10^2 nm [97, 99], biological processes within cells occur on much larger spatiotemporal scales. Therefore, new techniques have been developed not only to be able to simulate biological processes such as protein aggregation, translocation, cotranslational events, but also to obtain rigorous statistics on the energetic description of those processes and tools to analyze increasing amount of data. For instance, in the fast time scale and short distance region of the spatiotemporal landscape accessible to molecular dynamics, quantum mechanics processes are studied by using a quantum mechanic/molecular mechanic (QM/MM) approach. The entire biological system is divided into two sub-regions, one of which comprises the active site and it is treated quantum mechanically, while the other, comprising the rest of the system, is treated classically and simulated with atomistic simulations [100–102]. Exploring the high-value region of the spatiotemporal landscape of biomolecules and cell processes by means of atomistic simulations is not feasible because of the enormous computational effort, which would certainly require months or years of computer time, and because of the statistics. In fact, again because of the important role of noise, in order to get proper information of a biological process we need to perform a large number of experiments in silico. To overcome this problem, longer simulations have been carried out by "coarse graining" the system, i.e., by reducing the degrees of freedom and the components of a system grouping single components and defining appropriate force fields that take into account those reductions [99, 103–105]. Obviously, even if it is possible to map back the coarse grain approximation to atomistic resolution, molecular dynamics at atomistic resolution remain a suitable tool to attempt at mechanistically elucidating specific biological problems. Therefore the past decade has seen a huge development not only of the hardware with the construction of supercomputers capable of performing specific molecular simulations, like DESRES Anton and IBM's Summit, or of parallelization algorithm, which for instance has been exploited by the folding@home project, which has allowed to carry out parallel simulations on personal computers all over the world [106], but also of statistical methods to enhance sampling, accelerate simulations, or find the shorter energetic transition path between two conformational (meta)stable states (see for instance Refs. [35–45] in [107]).

Deeply rooted in statistical mechanics methods, AI and ML approaches are rapidly expanding into molecular dynamics of biological systems. Already employed largely as classification and predictive tools in chemistry, physics, and drug design [108–110], AI has entered slowly the fields of computational biochemistry and biophysics and is swiftly revolutionizing them. The last and maybe most remarkable accomplishment of ML application that has solved a 50-year-old problem such as the protein folding [111–116]. Proteins are responsible for most cellular biological processes and the amino acid sequence encodes all the information for a protein to fold and fulfill their functions, but predicting the correct folded 3D structure has been a field of investigation since the first protein structure was resolved [111–113]. Since the early days, the "protein-folding" problem and the related aspect involving the physics, chemistry, and biology of folding have intrigued the scientific community. Here, we want to focus on how computational science and specifically ML has played a primary role in shedding light on a potential solution of the problem. Protein functions are determined by its 3D structure, which is encoded in the amino acid sequence. Experiments for resolving protein structures are often difficult and expensive to carry out, so computer-based protein-structure prediction became such a crucial tool that in 1994 an event called Critical Assessment of protein Structure Prediction (CASP) was initiated [117]. Every other year the CASP community competes to predict the 3D structure of 100 selected target proteins, whose structures are known but not publicly available yet. Successful strategies for predicting 3D spatial protein conformation were based on fragment assembly, where a protein sequence is divided into small fragments, a search is conducted on the Protein Data Bank (PDB), and a full-length prediction is provided based on structures of sequence containing similar fragments [113]. However, in recent years ML methods combined with the use of evolutionary information have been proved to be crucial to improve the accuracy of structure predictions [118]. In order to preserve biological functions, the mutation of one amino acid is often accompanied by a mutation of the corresponding contacting amino acid and the co-evolution of mutations reveals information on conformational distances between residues and several methods including ultra-deep learning models and fully convolutional neural networks have been demonstrated to show outstanding performance in inferring contact maps and predicting distance matrices from co-evolutionary information [119–122]. Surprisingly, in 2018 Google's DeepMind presented at CASP13 the AlphaFold system [114, 115], which is heavily rooted in ML methods and ranked first in predicting protein structure. Remarkably, in the following CASP14 edition of 2020, AlphaFold 2 outperformed not only all the other teams participating in the contest, but also its previous version [116]. According to the scientific community, a 50-year-old problem is solved and this approach could generate a revolution in biology.

ML approaches are becoming common to analyze protein and macromolecule dynamics [123–126]. Shallow ML approaches such as clustering algorithms have been employed ever since the first simulations were carried out to identify different relevant conformational states and investigate the transitions among them [127–129]. Closely related to clustering methods are Markov state Models, which model the transition between free energy basins through a Markovian transition or rate matrix [130–132]. Other common use of shallow ML methods concerns the reduction of the dimensionality of the system in the form of principal component analysis or for exploring in a more efficient way the conformational energy landscape [126].

Nowadays, not only shallow ML methods but also deep AI approaches in general are crucial in advancing computational biophysics, biochemistry, and physical chemistry [134]. Both supervised and unsupervised learning are used to analyze data, make predictions, and explore new uncharted regions of the feature space of a system. For instance, for molecular systems as small as water, deep neural networks allowed the investigation of water anomalies [135, 136], enabled predictions of the hydration free energy of a cavity near a monolayer interface [137] and of surface charge density from interfacial water orientation [138]. In computer-aided drug-design, a variety of AI approaches have been used largely ever since also to predict drug sensitivity in cancer cells (see for instance references in [139]). The development of deep learning methods has great influence also on this field [140–142]. In line with these developments, a framework rooted in attention-based multimodal neural approach for predicting drug sensitivity for cancer cells has demonstrated to be excellent [140–142]. On the other hand, AI approaches also benefited predictions concerning large macromolecules such as chromatin, for instance, the complex systems in which DNA is folded and packed compact with the aid of histon proteins [143].

Overall, these methods and their applications to the various subfields of biophysics are crucial to their wider diffusion. Synthetic biology is still behind in the wider application of such methods, and that is necessarily dependent on the computational burden that historically has affected the field. The sheer acceleration provided by the use of AI algorithms is an important stepping stone for their general diffusion. More importantly, algorithms linking sequence to structure are still few. An important bottleneck is the lack of training data for much of the algorithms described in this section. To overcome this problem, transfer learning and generative models are usually employed in other fields. In the next section, we will provide a short overview of the field of generative methods and their possible application to synthetic biology.

7. Synthetic Design Using Generative Models

An exciting avenue of research in synthetic biology using AI is the adoption of generative models. These are computational methods which are capable of producing synthetic data which share general statistical features with real data. Application of these methods is still nascent in the field, with very few examples to be reported. However, their importance cannot be overestimated. In fact, generative models, when paired with algorithms of reinforcement learning or evolutionary algorithms, can provide design principles of biological structures in a completely unsupervised way. The importance of this statement lies in the observation that training an algorithm with existing data is bound to be biased. In this case, training an AI method with biological observations can bias the algorithm and lead it to propose designs that are not very different from existing ones. On the other hand, unsupervised learning reinforced using score functions that encode general and simple biological rules may provide designs outside what has been observed in 4 billion years of evolution.

Examples of such methods are the Generative Adversarial Networks (GANs) [144]. GANs consist of two neural networks engaged in a zero-sum adversarial game, where one network, the generator, is engaged in producing synthetic data, and the other, the discriminator, is posed to distinguish the synthetic data from the real data. A feedback mechanism ensures refinement of both models during the game, with the end goal of the generative network being able to produce data virtually indistinguishable by the discriminator network from the real data. While GANs have been used successfully in many applications, especially in image reconstruction and enhancement, their application in synthetic biology is very sparse. A recent example is the paper of Gupta and Zou [145], which uses a feedback GAN to suggest protein sequences associated with a desired function.

Another recent paper used a Monte Carlo Tree Search reinforcement learning algorithm guided by chemical similarity for retrosynthesis [146]. The algorithm was trained using the reaction rules describing bonding patterns when a set of substrates is transformed into a set of products, random sampling was used to choose the transformation. The transformation process was then simulated forward in time, providing a score using the biochemical rules. The final score was then back-propagated to update and rank the initial design choices.

In general, such models can be applied across scales. However, in the field of synthetic biology, the highest impact of generative models is predicted in the context of a unified design theory. Linked with mechanistic modeling and systems biology, molecular dynamics and biophysics, and cell to cell interaction, the use of generative models can produce never before seen structures with desired properties inferred from the other scales, a single model that would provide the mechanistic rules needed to build an AI capable of producing cell designs in a rational and repeatable way. While the details would be dependent on the modeling modules, a general strategy for structuring an AI to learn the biological rules might be as follows: The physical structure of the cell would be presented to the AI, together with the mechanistic rules beforehand. The AI will construct examples of cellular sensors by running in parallel all the computational modules, scoring against a target phenotype (e.g., chemical sensing) as a function of time. As time progresses, structures that underperform against an appropriate score function or predicted phenotype algorithm would be discarded, while those that are predicted to perform better would serve to "reinforce" the AI. Ultimately, an ideal set of structures would be output by the model, providing precise, rational, and testable cellular designs. Using a mechanistic model to train an AI using reinforcement learning is a task that has never been attempted. The multi-scale nature of this process is also an additional complication. By dividing the problem in several layers that can be solved independently, the high-dimensional problem of learning all of the modules at the same time would be reduced to the simpler but still non-trivial one of learning the design of a single module at the time.

8. Conclusions

In this chapter, we have reviewed some of the most recent advances in the field of AI for synthetic biology, with specific regards to the construction of cellular biosensors. Specifically, we have provided examples of how AI can be used to accelerate the investigation of the parameter space of a mechanistic model and molecular dynamic simulations, and how the use of generative models can provide completely new designs, outside what has been observed in Nature. The complete end-to-end realization of a computational platform for the design of biological systems is still impossible today. But luckily the tools to build it are already available today.

Acknowledgments

This material is based upon work supported by the National Science Foundation under Grant No. DBI-1548297. Disclaimer: Any opinions, findings, and conclusions or recommendations expressed in this material are those of the authors and do not necessarily reflect the views of the National Science Foundation.

References

[1] Moen, E., Bannon, D., Kudo, T., Graf, W., Covert, M., and Van Valen, D. (2019). Deep learning for cellular image analysis. *Nature methods* 16(12): 1233–1246.

[2] Nichols, J. A., Chan, H. W. H., and Baker, M. A. (2019). Machine learning: Applications of artificial intelligence to imaging and diagnosis. *Biophys. Rev.* 11(1): 111–118.

[3] Van Valen, D. A., Kudo, T., Lane, K. M., Macklin, D. N., Quach, N. T., DeFelice, M. M., Maayan, I., Tanouchi, Y., Ashley, E. A., and Covert, M. W. (2016). Deep learning automates the quantitative analysis of individual cells in live-cell imaging experiments. *PLOS Comput. Biol.* 12(11): e1005177, doi:10.1371/journal.pcbi.1005177, http://dx.plos.org/10.1371/journal.pcbi.1005177.

[4] Leung, M. K., Delong, A., Alipanahi, B., and Frey, B. J. (2015). Machine learning in genomic medicine: A review of computational problems and data sets. *Proc. IEEE* 104(1): 176–197.

[5] Yip, K. Y., Cheng, C., and Gerstein, M. (2013). Machine learning and genome annotation: a match meant to be?. *Genome biology* 14(5): 1–10.

[6] Alipanahi, B., Delong, A., Weirauch, M. T., and Frey, B. J. (2015). Predicting the sequence specificities of DNA- and RNA-binding proteins by deep learning. *Nat. Biotechnol.* 33(8): 831–838, doi:10.1038/nbt.3300. URL: http://dx.doi.org/10.1038/nbt.3300.

[7] Ballester, P. J., and Mitchell, J. B. (2010). A machine learning approach to predicting protein–ligand binding affinity with applications to molecular docking. *Bioinformatics* 26(9): 1169–1175.

[8] Califano, A., and Alvarez, M. J. (2017). The recurrent architecture of tumour initiation, progression and drug sensitivity. *Nat. Rev. Cancer* 17(2): 116.

[9] Carro, M. S., Lim, W. K., Alvarez, M. J., Bollo, R. J., Zhao, X., Snyder, E. Y., Sulman, E. P., Anne, S. L., Doetsch, F., Colman, H. *et al.* (2010). The transcriptional network for mesenchymal transformation of brain tumours. *Nature* 463(7279): 318–325.

[10] Langille, M. G., Zaneveld, J., Caporaso, J. G., McDonald, D., Knights, D., Reyes, J. A., Clemente, J. C., Burkepile, D. E., Thurber, R. L. V., Knight, R. *et al.* (2013). Predictive functional profiling of microbial communities using 16s rrna marker gene sequences. *Nat. Biotechnol.* 31(9): 814–821.

[11] Djebali, S., Davis, C. A., Merkel, A., Dobin, A., Lassmann, T., Mortazavi, A., Tanzer, A., Lagarde, J., Lin, W., Schlesinger, F. *et al.* (2012). Landscape of transcription in human cells. *Nature* 489(7414): 101–108.

[12] Marbach, D., Costello, J. C., Küffner, R., Vega, N. M., Prill, R. J., Camacho, D. M., Allison, K. R., Kellis, M., Collins, J. J., and Stolovitzky, G. (2012). Wisdom of crowds for robust gene network inference. *Nat. Methods* 9(8): 796–804.

[13] Valeri, J. A., Collins, K. M., Ramesh, P., Alcantar, M. A., Lepe, B. A., Lu, T. K., and Camacho, D. M. (2020). Sequence-to-function deep learning frameworks for engineered riboregulators. *Nat. Commun.* 11(1): 1–14.

[14] Wang, J., Cao, H., Zhang, J. Z., and Qi, Y. (2018). Computational protein design with deep learning neural networks. *Sci. Rep.* 8(1): 1–9.

[15] Radivojević, T., Costello, Z., Workman, K., and Garcia Martin, H. (2020). A machine learning Automated Recommendation Tool for synthetic biology. *Nat. Commun.* 11(1): 1–14. arXiv:1911.11091, doi:10.1038/s41467-020-18008-4.

[16] Zhang, J., Petersen, S. D., Radivojevic, T., Ramirez, A., Pérez-Manríquez, A., Abeliuk, E., Sánchez, B. J., Costello, Z., Chen, Y., Fero, M. J. *et al.* (2020). Combining mechanistic and machine learning models for predictive engineering and optimization of tryptophan metabolism. *Nat. Commun.* 11(1): 1–13.

[17] Chen, B., Khodadoust, M. S., Olsson, N., Wagar, L. E., Fast, E., Liu, C. L., Muftuoglu, Y., Sworder, B. J., Diehn, M., Levy, R., Davis, M. M., Elias, J. E., Altman, R. B., and Alizadeh, A. A. (2019). Predicting HLA class II

antigen presentation through integrated deep learning. *Nat. Biotechnol.* 37(11): 1332–1343, doi:10.1038/s41587-019-0280-2.

[18] Racle, J., Michaux, J., Rockinger, G. A., Arnaud, M., Bobisse, S., Chong, C., Guillaume, P., Coukos, G., Harari, A., Jandus, C., Bassani-Sternberg, M., and Gfeller, D. (2019). Robust prediction of HLA class II epitopes by deep motif deconvolution of immunopeptidomes. *Nat. Biotechnol.* 37(11): 1283–1286, doi:10.1038/s41587-019-0289-6, http://dx.doi.org/10.1038/s41587-019-0289-6.

[19] Nielsen, J., and Keasling, J. D. (2016). Engineering cellular metabolism. *Cell* 164(6): 1185–1197.

[20] Ajikumar, P. K., Xiao, W.-H., Tyo, K. E., Wang, Y., Simeon, F., Leonard, E., Mucha, O., Phon, T. H., Pfeifer, B., and Stephanopoulos, G. (2010). Isoprenoid pathway optimization for taxol precursor overproduction in escherichia coli. *Science* 330(6000): 70–74.

[21] Galanie, S., Thodey, K., Trenchard, I. J., Interrante, M. F., and Smolke, C. D. (2015). Complete biosynthesis of opioids in yeast. *Science* 349(6252): 1095–1100.

[22] Paddon, C. J., Westfall, P. J., Pitera, D. J., Benjamin, K., Fisher, K., McPhee, D., Leavell, M., Tai, A., Main, A., D. Eng *et al.* (2013). High-level semi-synthetic production of the potent antimalarial artemisinin. *Nature* 496(7446): 528–532.

[23] Eichenberger, M., Lehka, B. J., Folly, C., Fischer, D., Martens, S., Simón, E., and Naesby, M. (2017). Metabolic engineering of saccharomyces cerevisiae for de novo production of dihydrochalcones with known antioxidant, antidiabetic, and sweet tasting properties. *Metab. Eng.* 39: 80–89.

[24] Atsumi, S., Hanai, T., and Liao, J. C. (2008). Non-fermentative pathways for synthesis of branched-chain higher alcohols as biofuels. *Nature* 451(7174): 86–89.

[25] Yang, X., Xu, M., and Yang, S.-T. (2015). Metabolic and process engineering of clostridium cellulovorans for biofuel production from cellulose. *Metab. Eng.* 32: 39–48.

[26] Hodgman, C. E., and Jewett, M. C. (2012). Cell-free synthetic biology: Thinking outside the cell. *Metab. Eng.* 14(3): 261–269.

[27] Zhao, W., Schafer, S., Choi, J., Yamanaka, Y. J., Lombardi, M. L., Bose, S., Carlson, A. L., Phillips, J. A., Teo, W., Droujinine, I. A., Cui, C. H., Jain, R. K., Lammerding, J., Love, J. C., Lin, C. P., Sarkar, D., Karnik, R., and Karp, J. M. (2011). Cell-surface sensors for real-time probing of cellular environments. *Nat. Nanotechnol.* 6(8): 524–531, doi:10.1038/nnano.2011.101.

[28] Qian, L., Li, Q., Baryeh, K., Qiu, W., Li, K., Zhang, J., Yu, Q., Xu, D., Liu, W., Brand, R. E. *et al.* (2019). Biosensors for early diagnosis of pancreatic cancer: A review. *Transl. Res.* 213: 67–89.

[29] Cui, F., Yue, Y., Zhang, Y., Zhang, Z., and Zhou, H. S. (2020). Advancing biosensors with machine learning. *ACS sensors* 5(11): 3346–3364.
[30] Cui, F., Zhou, Z., Feng, H., and Susan Zhou, H. (2020). Disposable polyurethane nanospiked gold electrode-based label-free electrochemical immunosensor for clostridium difficile. *ACS Appl. Nano Mater.* 3(1): 357–363, doi:10.1021/acsanm.9b02001.
[31] Richardson, S. M., Mitchell, L. A., Stracquadanio, G., Yang, K., Dymond, J. S., DiCarlo, J. E., Lee, D., Huang, C. L. V., Chandrasegaran, S., Cai, Y., Boeke, J. D., and Bader, J. S. (2017). Design of a synthetic yeast genome. *Science* 355(6329): 1040–1044, doi:10.1126/science.aaf4557. URL: http://science.sciencemag.org/content/355/6329/1040.
[32] Deamer, D. (2005). A giant step towards artificial life? *Trends in biotechnology* 23(7): 336–338.
[33] Nielsen, A. A., Der, B. S., Shin, J., Vaidyanathan, P., Paralanov, V., Strychalski, E. A., ... and Voigt, C. A. (2016). Genetic circuit design automation. *Science* 352(6281).
[34] Donahue, P. S., Draut, J. W., Muldoon, J. J., Edelstein, H. I., Bagheri, N., and Leonard, J. N. (2020). The COMET toolkit for composing customizable genetic programs in mammalian cells. *Nat. Commun.* 11(1), doi:10.1038/s41467-019-14147-5. URL: http://dx.doi.org/10.1038/s41467-019-14147-5.
[35] Wiesenfeld, K., and Moss, F. (1995). Stochastic resonance and the benefits of noise: From ice ages to crayfish and squids. *Nature* 373(6509): 33–36.
[36] Rao, C. V., Wolf, D. M., and Arkin, A. P. (2002). Control, exploitation and tolerance of intracellular noise. *Nature* 420(6912): 231–237.
[37] Elowitz, M. B., Levine, A. J., Siggia, E. D., and Swain, P. S. (2002). Stochastic gene expression in a single cell. *Science* 297(5584): 1183–1186.
[38] Casanova, M. P. (2020). Noise and synthetic biology: How to deal with stochasticity? *NanoEthics* 14(1): 113–122.
[39] Kittisopikul, M., and Süel, G. M. (2010). Biological role of noise encoded in a genetic network motif. *Proc. Natl. Acad. Sci.* 107(30): 13300–13305.
[40] Ramaswamy, R., and Sbalzarini, I. F. (2011). Intrinsic noise alters the frequency spectrum of mesoscopic oscillatory chemical reaction systems. *Sci. Rep.* 1: 154.
[41] Wang, S., and Raghavachari, S. (2011). Quantifying negative feedback regulation by micro-rnas. *Phys. Biol.* 8(5): 055002.
[42] Balázsi, G., van Oudenaarden, A., and Collins, J. J. (2011). Cellular decision making and biological noise: From microbes to mammals. *Cell* 144(6): 910–925.
[43] Dobzhansky, T. (1973). Nothing in biology makes sense except in the light of evolution. *Am. Biol. Teach.* 35(3): 125–129.

[44] Endy, D. (2005). Foundations for engineering biology. *Nature* 438(7067): 449–453, doi:10.1038/nature04342.

[45] Di Natale, F., Bhatia, H., Carpenter, T. S., Neale, C., Kokkila-Schumacher, S., Oppelstrup, T., ... and Ingólfsson, H. I. (2019, November). A massively parallel infrastructure for adaptive multiscale simulations: Modeling RAS initiation pathway for cancer. In *Proceedings of the International Conference for High Performance Computing, Networking, Storage and Analysis*, pp. 1–16.

[46] Karr, J. R., Sanghvi, J. C., MacKlin, D. N., Gutschow, M. V., Jacobs, J. M., Bolival, B., Assad-Garcia, N., Glass, J. I., and Covert, M. W. (2012). A whole-cell computational model predicts phenotype from genotype. *Cell* 150(2): 389–401, doi:10.1016/j.cell.2012.05.044.

[47] Bianco, S., Chan, Y.-H. M., and Marshall, W. F. (2020). Towards computer-aided design of cellular structure. *Phys. Biol.* 17(2): 023001.

[48] Marques, M. E., Mansur, A. A., and Mansur, H. S. (2013). Chemical functionalization of surfaces for building three-dimensional engineered biosensors. *Appl. Surf. Sci.* 275: 347–360.

[49] Park, M., Tsai, S.-L., and Chen, W. (2013). Microbial biosensors: Engineered microorganisms as the sensing machinery. *Sensors* 13(5): 5777–5795.

[50] Campàs, M., Prieto-Simón, B., and Marty, J. L. (2009, February). A review of the use of genetically engineered enzymes in electrochemical biosensors. In *Seminars in Cell & Developmental Biology*, Academic Press, 20(1): pp. 3–9.

[51] Leth, S., Maltoni, S., Simkus, R., Mattiasson, B., Corbisier, P., Klimant, I., Wolfbeis, O. S., and Csöregi, E. (2002). Engineered bacteria based biosensors for monitoring bioavailable heavy metals. *Electroanalysis: An International Journal Devoted to Fundamental and Practical Aspects of Electroanalysis* 14(1): 35–42.

[52] Shin, H. J. (2011). Genetically engineered microbial biosensors for in situ monitoring of environmental pollution. *Appl. microbiol. Biotechnol.* 89(4): 867–877.

[53] Toda, S., Blauch, L. R., Tang, S. K., Morsut, L., and Lim, W. A. (2018). Programming self-organizing multicellular structures with synthetic cell-cell signaling. *Science* 361(6398): 156–162, doi:10.1126/science.aat0271.

[54] Barrett, D. M., Singh, N., Porter, D. L., Grupp, S. A., and June, C. H. (2014). Chimeric antigen receptor therapy for cancer. *Ann. Rev. Med.* 65: 333–347.

[55] Miller, J. F. and Sadelain, M. (2015). The journey from discoveries in fundamental immunology to cancer immunotherapy. *Cancer Cell* 27(4): 439–449.

[56] Sadelain, M., Brentjens, R., and Rivière, I. (2009). The promise and potential pitfalls of chimeric antigen receptors. *Curr. Opin. Immunol.* 21(2): 215–223.

[57] Morgan, R. A., Yang, J. C., Kitano, M., Dudley, M. E., Laurencot, C. M., and Rosenberg, S. A. (2010). Case report of a serious adverse event following the administration of t cells transduced with a chimeric antigen receptor recognizing erbb2. *Mol. Ther.* 18(4): 843–851.

[58] Slomovic, S., Pardee, K., and Collins, J. J. (2015). Synthetic biology devices for in vitro and in vivo diagnostics. *Proc. Natl. Acad. Sci.* 112(47): 14429–14435.

[59] Kopan, R. (2002). Notch: A membrane-bound transcription factor. *J. Cell Sci.* 115 (6): 1095–1097.

[60] Morsut, L., Roybal, K. T., Xiong, X., Gordley, R. M., Coyle, S. M., Thomson, M., and Lim, W. A. (2016). Engineering customized cell sensing and response behaviors using synthetic notch receptors. *Cell* 164(4): 780–791.

[61] Roybal, K. T., Rupp, L. J., Morsut, L., Walker, W. J., McNally, K. A., Park, J. S., and Lim, W. A. (2016). Precision tumor recognition by t cells with combinatorial antigen-sensing circuits. *Cell* 164(4): 770–779.

[62] Roybal, K. T., Williams, J. Z., Morsut, L., Rupp, L. J., Kolinko, I., Choe, J. H., Walker, W. J., Mc-Nally, K. A. and Lim, W. A. (2016). Engineering t cells with customized therapeutic response programs using synthetic notch receptors. *Cell* 167(2): 419–432.

[63] Roybal, K. T. and Lim, W. A. (2017). Synthetic immunology: Hacking immune cells to expand their therapeutic capabilities. *Ann. Rev. Immunol.* 35: 229–253.

[64] Yuting Z. and Ganesh S. (2010). Mathematical modeling: Bridging the gap between concept and realization in synthetic biology. *Bio. Med. International*, Article ID 541609, 2010: 16, https://doi.org/10.1155/2010/541609.

[65] Chakraborty, A. K. and Das, J. (2010). Pairing computation with experimentation: A powerful coupling for understanding T cell signaling. *Nat. Rev. Immunol.* 10(1): 59–71.

[66] Molina-París, C. and Lythe, G. (eds.). (2011). Mathematical models and immune cell biology. Springer.

[67] Gottschalk, R. A., Martins, A. J., Sjoelund, V. H., Angermann, B. R., Lin, B., and Germain, R. N. (2013, October). Recent progress using systems biology approaches to better understand molecular mechanisms of immunity. In *Seminars in immunology*, Academic Press, 25(3): pp. 201–208.

[68] Lever, M., Maini, P. K., Van Der Merwe, P. A. and Dushek, O. (2014). Phenotypic models of t cell activation. *Nat. Rev. Immunol.* 14(9): 619–629.

[69] Lopatkin, A. J. and Collins, J. J. (2020). Predictive biology: Modelling, understanding and harnessing microbial complexity. *Nature Reviews Microbiology* 18(9): 507–520.

[70] Gardner, T. S., Cantor, C. R., and Collins, J. J. (2000). Construction of a genetic toggle switch in escherichia coli. *Nature* 403(6767): 339–342.

[71] Elowitz, M. B. and Leibler, S. (2000). A synthetic oscillatory network of transcriptional regulators. *Nature* 403(6767): 335–338.

[72] Ferrell Jr., J. E. (2002). Self-perpetuating states in signal transduction: Positive feedback, double-negative feedback and bistability. *Curr. Opin. Cell Biol.* 14(2): 140–148.

[73] Kramer, B. P. and Fussenegger, M. (2005). Hysteresis in a synthetic mammalian gene network. Proc. *Natl. Acad. Sci.* 102(27): 9517–9522.

[74] Rosenfeld, N., Elowitz, M. B., and Alon, U. (2002). Negative autoregulation speeds the response times of transcription networks. *J. Mol. Biol.* 323(5): 785–793.

[75] Becskei, A. and Serrano, L. (2000). Engineering stability in gene networks by autoregulation. *Nature* 405(6786): 590–593.

[76] Dublanche, Y., Michalodimitrakis, K., Kümmerer, N., Foglierini, M., and Serrano, L. (2006). Noise in transcription negative feedback loops: Simulation and experimental analysis. *Mol. Syst. Biol.* 2(1): 41.

[77] Ferguson, M. L., Le Coq, D., Jules, M., Aymerich, S., Radulescu, O., Declerck, N., and Royer, C. A. (2012). Reconciling molecular regulatory mechanisms with noise patterns of bacterial metabolic promoters in induced and repressed states. *Proc. Natl. Acad. Sci.* 109(1): 155–160.

[78] Marquez-Lago, T. T. and Stelling, J. (2010). Counter-intuitive stochastic behavior of simple gene circuits with negative feedback. *Biophys. J.* 98(9): 1742–1750.

[79] Shimoga, V., White, J. T., Li, Y., Sontag, E., and Bleris, L. (2013). Synthetic mammalian transgene negative autoregulation. *Mol. Syst. Biol.* 9(1): 670.

[80] Novák, B. and Tyson, J. J. (2008). Design principles of biochemical oscillators. *Nat. Rev. Mol. Cell Biol.* 9(12): 981–991.

[81] Fung, E., Wong, W. W., Suen, J. K., Bulter, T., Lee, S.-g. and Liao, J. C. (2005). A synthetic gene–metabolic oscillator. *Nature* 435(7038): 118–122.

[82] Chilov, D. and Fussenegger, M. (2004). Toward construction of a self-sustained clock-like expression system based on the mammalian circadian clock. *Biotechnol. Bioeng.* 87(2): 234–242.

[83] Wang, S., Fan, K., Luo, N., Cao, Y., Wu, F., Zhang, C., Heller, K. A., and You, L. (2019). Massive computational acceleration by using neural networks to emulate mechanism-based biological models. *Nat. Commun.* 10(1): 1–9.

[84] Heckmann, D., Lloyd, C. J., Mih, N., Ha, Y., Zielinski, D. C., Haiman, Z. B., Desouki, A. A., Lercher, M. J., and Palsson, B. O. (2018). Machine learning applied to enzyme turnover numbers reveals protein structural correlates and improves metabolic models. *Nat. Commun.* 9(1): 1–10.

[85] Hiscock, T. W. (2019). Adapting machine-learning algorithms to design gene circuits. *BMC Bioinform.* 20(1): 1–13.

[86] Yang, J. H., Wright, S. N., Hamblin, M., McCloskey, D., Alcantar, M. A., Schrübbers, L., Lopatkin, A. J., Satish, S., Nili, A., Palsson B. O., *et al.* (2019). A white-box machine learning approach for revealing antibiotic mechanisms of action. *Cell* 177(6): 1649–1661.

[87] Cao, Y., Ryser, M. D., Payne, S., Li, B., Rao, C. V., and You, L. (2016). Collective space-sensing coordinates pattern scaling in engineered bacteria. *Cell* 165(3): 620–630.

[88] Pastore, V. P., Zimmerman, T., Biswas, S. K., and Bianco, S. (2019). Establishing the baseline for using plankton as biosensor. In *Imaging, Manipulation, and Analysis of Biomolecules, Cells, and Tissues XVII, SPIE*, Farkas, D. L., Leary, J. F., and Tarnok, A. (eds.), p. 16, doi:10.1117/12.2511065. URL: https://www.spiedigitallibrary.org/conference-proceedings-of-spie/10881/2511065/Establishing.

[89] Pastore, V. P., Zimmerman, T., Bisvas, S. K., and Bianco, S. (2020). Monitoring water quality using plankton as biosensor. *JDREAM. J. Interdiscip. Res. Appl. Med.* 4(1): 15–20.

[90] Biswas, S. K., Zimmerman, T., Maini, L., Adebiyi, A., Bozano, L., Brown, C., Pastore, V. P., and Bianco, S. (2019). High throughput analysis of plankton morphology and dynamic. In *Imaging, Manipulation, and Analysis of Biomolecules, Cells, and Tissues XVII, SPIE*, Farkas, D. L., Leary, J. F., and Tarnok, A. (eds.), p. 8, doi:10.1117/12.2509168. URL: https://www.spiedigitallibrary.org/conference-proceedings-of-spie/10881/2509168/High-through.

[91] Zimmerman, T., Antipa, N., Elnatan, D., Murru, A., Biswas, S., Pastore, V., Bonani, M., Waller, L., Fung, J., Fenu, G., and Bianco, S. (2019). Stereo in-line holographic digital microscope. In *Three-Dimensional and Multidimensional Microscopy: Image Acquisition and Processing XXVI, SPIE*, Brown, T. G. and Wilson, T. (eds.), p. 38, doi:10.1117/12.2509033. URL: https://www.spiedigitallibrary.org/conference-proceedings-of-spie/10883/2509033/Stereo-in-li.

[92] Pastore, V. P., Zimmerman, T. G., Biswas, S. K., and Bianco, S. (2020). Annotation-free learning of plankton for classification and anomaly detection. *Sci. Rep.* 10(1): 1–15, doi:10.1038/s41598-020-68662-3. URL: https://doi.org/10.1038/s41598-020-68662-3.

[93] McCammon, J. A., Gelin, B. R., and Karplus, M. (1977). Dynamics of folded proteins. *Nature* 267(5612): 585–590.

[94] Rahman, A. (1964). Correlations in the motion of atoms in liquid argon. *Phys. Rev.* 136: A405–A411, doi:10.1103/PhysRev.136.A405. URL: https://link.aps.org/doi/10.1103/PhysRev.136.A405.

[95] Allen, M. P. and Tildesley, D. J. (1987). *Computer Simulations of Liquids.* Oxford: Clarendon Press.

[96] Frenkel, D. and Smit, B. (1996). *Understanding Molecular Simulation: From Algorithms to Applications.* San Diego: Academic Press.

[97] Dror, R. O., Dirks, R. M., Grossman, J. P., Xu, H., and Shaw, D. E. (2012). Biomolecular simulation: A computational microscope for molecular biology. *Ann. Rev. Biophys.* 41: 429–452.

[98] Karplus, M. and McCammon, J. A. (2002). Molecular dynamics simulations of biomolecules. *Nat. Struct. Mol. Biol.* 9: 646–652.

[99] Takada, S. (2012). Coarse-grained molecular simulations of large biomolecules. *Curr. Opin. Struct. Biol.* 22(2): 130–137, Theory and Simulation/Macromolecular Assemblages, doi:https://doi.org/10.1016/j.sbi. 2012.01.010, http://www.sciencedirect.com/science/article/pii/S0959440 X12000280.

[100] Warshel, A. and Levitt, M. (1976). Theoretical studies of enzymic reactions: dielectric, electrostatic and steric stabilization of the carbonium ion in the reaction of lysozyme. *J. Mol. Biol.* 103(2): 227–249.

[101] Friesner, R. A. and Guallar, V. (2005). Ab initio quantum chemical and mixed quantum mechanics/molecular mechanics (qm/mm) methods for studying enzymatic catalysis. *Annu. Rev. Phys. Chem.* 56: 389–427.

[102] Senn, H. M. and Thiel, W. (2009). Qm/mm methods for biolmolecular systems. *Angew. Chem., Int. Ed.* 48(7): 1198–229.

[103] Saunders, M. G., and Voth, G. A. (2013). Coarse-graining methods for computational biology. *Annual Review of Biophysics*, 42, 73–93.

[104] Clementi, C. (2008). Coarse-grained models of protein folding: Toy models or predictive tools? *Curr. Opin. Struct. Biol.* 18(1): 10–15.

[105] Trovato, F. and O'Brien, E. P. (2016). Insights into cotranslational nascent protein behavior from computer simulations. *Ann. Rev. Biophys.* 45: 345–369.

[106] Folding at home, https://foldingathome.org/.

[107] Bottaro, S. and Lindorff-Larsen, K. (2018). Biophysical experiments and biomolecular simulations: A perfect match? *Science* 361(6400): 355–360.

[108] Sumpter, B. G., Getino, C., and Noid, D. W. (1994). Theory and applications of neural computing in chemical science. *Ann. Rev. Phys. Chem.* 45(1): 439–481.

[109] Zupan, J., & Gasteiger, J. (1993). Neural networks for chemists: An introduction. VCH publishers.

[110] Zupan, J., & Gasteiger, J. (1999). Neural networks in chemistry and drug design. John Wiley & Sons, Inc.

[111] Chan, H. S. and Dill, K. A. (1993). The protein folding problem. *Phys. Today* 46(2): 24–32.

[112] Dill, K. A., Ozkan, S. B., Shell, M. S., and Weikl, T. R. (2008). The protein folding problem. *Annu. Rev. Biophys.* 37: 289–316.

[113] Dill, K. A. and MacCallum, J. L. (2012). The protein-folding problem, 50 years on. *Science* 338(6110): 1042–1046.

[114] AlQuraishi, M. (2019). Alphafold at casp13. *Bioinformatics* 35(22): 4862–4865.

[115] Senior, A. W., Evans, R., Jumper, J., Kirkpatrick, J., Sifre, L., Green, T., Qin, C., Zˇídek, A., Nelson, A. W., Bridgland, A. *et al.* (2020). Improved protein structure prediction using potentials from deep learning. *Nature* 577(7792): 706–710.

[116] Callaway, E. (2020). "It will change everything": AI makes gigantic leap in solving protein structures. *Nature* 588: 203–204.

[117] Moult, J. (2005). A decade of casp: Progress, bottlenecks and prognosis in protein structure prediction. *Curr. Opin. Struct. Biol.* 15(3): 285–289.

[118] Noé, F., De Fabritiis, G., and Clementi, C. (2020). Machine learning for protein folding and dynamics. *Curr. Opin. Struct. Biol.* 60: 77–84.

[119] Ma, J., Wang, S., Wang, Z., and Xu, J. (2015). Protein contact prediction by integrating joint evolutionary coupling analysis and supervised learning. *Bioinformatics* 31(21): 3506–3513.

[120] Wang, S., Sun, S., Li, Z., Zhang, R., and Xu, J. (2017). Accurate de novo prediction of protein contact map by ultra-deep learning model. *PLoS Comput. Biol.* 13(1): e1005324.

[121] Xu, J. (2019). Distance-based protein folding powered by deep learning. *Proc. Natl. Acad. Sci.* 116(34): 16856–16865.

[122] Jones, D. T. and Kandathil, S. M. (2018). High precision in protein contact prediction using fully convolutional neural networks and minimal sequence features. *Bioinformatics* 34(19): 3308–3315.

[123] Rohrdanz, M. A., Zheng, W., and Clementi, C. (2013). Discovering mountain passes via torchlight: Methods for the definition of reaction coordinates and pathways in complex macromolecular reactions. *Ann. Rev. Phys. Chem.* 64(1): 295–316.

[124] Von Lilienfeld, O. A. (2018). Quantum machine learning in chemical compound space. *Angew. Chem. Int. Ed.* 57(16): 4164–4169.

[125] Butler, K. T., Davies, D. W., Cartwright, H., Isayev, O., and Walsh, A. (2018). Machine learning for molecular and materials science. *Nature* 559(7715): 547–555.

[126] Ceriotti, M. (2019). Unsupervised machine learning in atomistic simulations, between predictions and understanding. *J. Chem. Phys.* 150(15): 150901.

[127] Levitt, M. (1983). Molecular dynamics of native protein. II. Analysis, and nature of motion. *J. Mol. Biol.* 168(3): 621–657.

[128] Karpen, M. E., Tobias, D. J., and Brooks III, C. L. (1993). Statistical clustering techniques for the analysis of long molecular dynamics trajectories: Analysis of 2.2-ns trajectories of ypgdv. *Biochemistry* 32(2): 412–420.

[129] Torda, A. E. and van Gunsteren, W. F. (1994). Algorithms for clustering molecular dynamics configurations. *J. Comput. Chem.* 15(12): 1331–1340.

[130] Pande, V. S., Beauchamp, K., and Bowman, G. R. (2010). Everything you wanted to know about markov state models but were afraid to ask. *Methods* 52(1): 99–105.

[131] Prinz, J.-H., Wu, H., Sarich, M., Keller, B., Senne, M., Held, M., Chodera, J. D., Schütte, C., and Noé, F. (2011). Markov models of molecular kinetics: Generation and validation. *J. Chem. Phys.* 134(17): 174105.

[132] Chodera, J. D. and Noé, F. (2014). Markov state models of biomolecular conformational dynamics. *Curr. Opin. Struct. Biol.* 25: 135–144.

[133] Ceriotti, M. (2019). Unsupervised machine learning in atomistic simulations, between predictions and understanding. *The Journal of Chemical Physics*, 150(15): 150901.

[134] Prezhdo, O. V. (2020). Advancing physical chemistry with machine learning. *J. Phys. Chem. Lett.* 1(22): 9656–9658.

[135] Martelli, F., Leoni, F., Sciortino, F., and Russo, J. (2020). Connection between liquid and non-crystalline solid phases in water. *J. Chem. Phys.* 153(10): 104503.

[136] Gartner, T. E., Zhang, L., Piaggi, P. M., Car, R., Panagiotopoulos, A. Z., and Debenedetti, P. G. (2020). Signatures of a liquid–liquid transition in an ab initio deep neural network model for water. *Proc. Natl. Acad. Sci.* 117(42): 26040–26046.

[137] Kelkar, A. S., Dallin, B. C., and Van Lehn, R. C. (2020). Predicting hydrophobicity by learning spatiotemporal features of interfacial water structure: Combining molecular dynamics simulations with convolutional neural networks. *J. Phys. Chem. B* 124(41): 9103–9114.

[138] Oh, M. I., Oh, C. I., Gupta, M., and Weaver, D. F. (2020). Decoding interfacial water orientation to predict surface charge density on a model sheet using a deep learning algorithm. *J. Phys. Chem. C* 124(4): 2574–2582.

[139] Manica, M., Oskooei, A., Born, J., Subramanian, V., Sáez-Rodríguez, J., and Rodríguez Martínez, M. (2019). Toward explainable anticancer compound sensitivity prediction via multimodal attention-based convolutional encoders. *Molecular Pharmaceutics* 16(12): 4797–4806.

[140] Chen, H., Engkvist, O., Wang, Y., Olivecrona, M., and Blaschke, T. (2018). The rise of deep learning in drug discovery. *Drug Discov. Today* 23(6): 1241–1250.

[141] Grapov, D., Fahrmann, J., Wanichthanarak, K., and Khoomrung, S. (2018). Rise of deep learning for genomic, proteomic, and metabolomic

data integration in precision medicine. *Omics: J. Integr. Biol.* 22(10): 630–636.

[142] Wu, Z., Ramsundar, B., Feinberg, E. N., Gomes, J., Geniesse, C., Pappu, A. S., Leswing, K., and Pande, V. (2018). Moleculenet: A benchmark for molecular machine learning. *Chem. Sci.* 9(2): 513–530.

[143] Zhang, R. and Ma, J. (2020). Matcha: Probing multi-way chromatin interaction with hypergraph representation learning. *Cell Syst.* 10(5): 397–407.

[144] Goodfellow, I., Pouget-Abadie, J., Mirza, M., Xu, B., Warde-Farley, D., Ozair, S., ... and Bengio, Y. (2014). Generative adversarial nets. Advances in neural information processing systems (NIPS), 27.

[145] Gupta, A. and Zou, J. (2019). Feedback gan for dna optimizes protein functions. *Nat. Mach. Intell.* 1(2): 105–111.

[146] Koch, M., Duigou, T., and Faulon, J. L. (2019). Reinforcement Learning for Bio-Retrosynthesis, doi:10.1101/800474. URL: https://uk.mathworks.com/videos/reinforcement-learning-part-1-what-is-reinforcement-learning.

Index